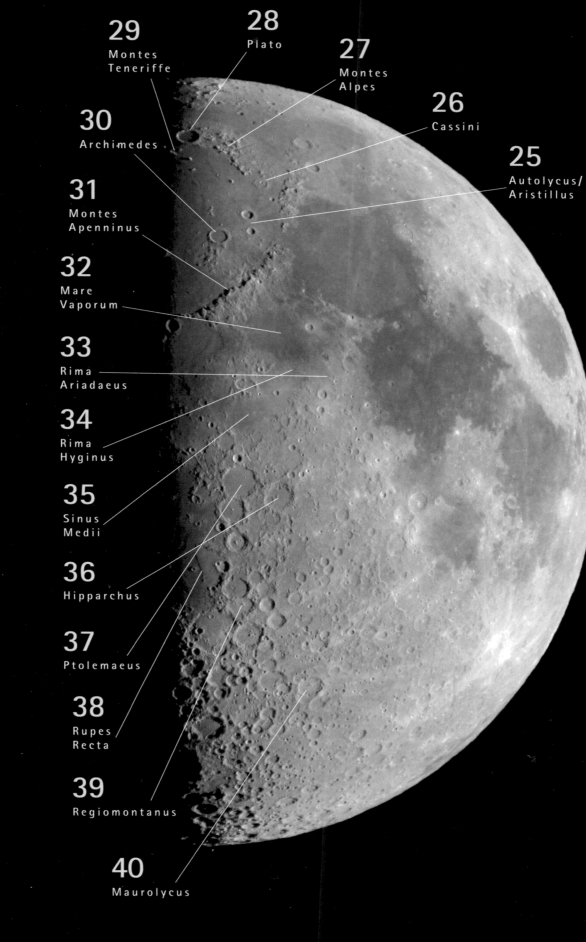

29 Montes Teneriffe

28 Plato

27 Montes Alpes

26 Cassini

25 Autolycus/ Aristillus

30 Archimedes

31 Montes Apenninus

32 Mare Vaporum

33 Rima Ariadaeus

34 Rima Hyginus

35 Sinus Medii

36 Hipparchus

37 Ptolemaeus

38 Rupes Recta

39 Regiomontanus

40 Maurolycus

21 osidonius

19 Mare Tranquillitatis

17 Sinus Asperitatis

16 Theophilus

13 Mare Nectaris

The Cambridge Photographic
Moon Atlas

Using the latest methods in digital photography and image processing, *The Cambridge Photographic Moon Atlas* presents sixty-nine regions of the lunar landscape in large-format images with corresponding charts. Each two-page spread features a specific region in multiple lighting situations, allowing for direct comparisons of the same landscape at different lunar phases. Detailed descriptions of each region's location and topography accompany 388 high-resolution photographs, making the Moon's craters, mountains, valleys, rilles, and other features easy to identify. Tracing the morphology of the Moon in unprecedented detail, this comprehensive and accessible visual atlas is an indispensable aid for amateur astronomers, astrophotographers, and casual observers.

ALAN CHU is an astrophotographer based in Hong Kong. Author of *The Photographic Moon Book* (2004) and *Foundations of Astronomy* (2010), he has also published several articles for the Hong Kong Astronomical Society, of which he is a founding member.

WOLFGANG PAECH is the former director of the Public Observatory in Hannover, Germany. He has published several books for amateur astronomers, including *Tips and Tricks for Amateur Astronomers* and *The Solar Handbook*.

MARIO WEIGAND is a Ph.D. candidate in nuclear physics at Goethe University in Frankfurt, Germany. He has eighteen years experience as an amateur astronomer and astrophotographer.

STORM DUNLOP is an experienced writer and lecturer on astronomy. He is author of *Collins Night Sky* (2011), a Fellow of the Royal Astronomical Society, and a past president of the British Astronomical Association.

CAMBRIDGE UNIVERSITY PRESS

Cambridge, New York, Melbourne, Madrid, Cape Town,
Singapore, São Paulo, Delhi, Mexico City

Cambridge University Press
32 Avenue of the Americas, New York, NY 10013, USA

www.cambridge.org
Information on this title: www.cambridge.org/9781107019737

English edition © Cambridge University Press 2012
© Oculum-Verlag GmbH, Erlangen 2010

First published in German by Oculum-Verlag GmbH, Erlangen, 2010.
English edition published 2012.

Printed in China by Everbest.

A catalogue record for this publication is available from the British Library.

ISBN   978-1-107-01973-7   Hardback

The Cambridge Photographic
Moon Atlas

Alan Chu, Wolfgang Paech, Mario Weigand

Translated by Storm Dunlop

Preface

Photographic observation of the Moon is undergoing a worldwide renaissance among present-day amateurs, partly through professional video cameras at affordable prices and even more through the availability of image-processing software written by amateurs. With modern imaging techniques and the corresponding software for image processing and observational planning, amateurs today obtain images of the Moon, taken with telescopes with apertures of 150–350 mm, in which the resolution is in the sub-arcsecond region. Their quality is far superior to the professional images in the large lunar atlases of the last century. The 388 Moon photographs shown in this lunar atlas were chosen from the resources of the three participating astrophotographers. They were all obtained with digital cameras. Alan Chu, a lunar photographer from Hong Kong, mainly uses a 10-inch Newtonian. Mario Weigand, an astrophotographer from Offenbach, took his photographs with 11- and 14-inch Schmidt-Cassegrain telescopes, while Wolfgang Paech from Hannover photographed the Moon with a 6-inch refractor and a 14-inch Schmidt-Cassegrain, not only from Germany, but also from Namibia. The photographs from the three authors were augmented by individual photos by Michael Theusner and Wolfgang Sorgenfrey.

The basis for the text of this book came from the *Photographic Moon Book* by Alan Chu. It was translated, brought up-to-date, and significantly expanded by Wolfgang Paech.

The authors thank Wolfgang Sorgenfrey, a pioneer of high-resolution lunar photography in Germany. He made his best images available and initiated many stimulating email discussions on the subject of image processing. Thanks are also due to Michael Theusner, the programmer of AviStack, who provided some high-resolution colour images. Martin Rietze placed images of terrestrial examples at our disposal. Mardiña Clark kindly gave permission for the resolution test charts to be reproduced, and Wilfried Tost gave decisive help in proof-reading. Finally, thanks go to Franz Hofmann, who works in the field of Lunar Laser Ranging (LLR), and who contributed significant information about the subject.

Advice on using this book

Hugh John Gramatzki, a widely known amateur astronomer, physicist and publicist of the last century, coined the phrase 'Every telescope has its own sky'. And that still applies in the 21st century: 'Every telescope has its own Moon'. For every size of telescope, beginning with good binoculars or a small telescope, there is a rewarding object to observe. This book should stimulate readers to undertake their own – visual or photographic – observations.

The short introductory text section gives a summary of the current state of knowledge and explains the important relationships that are necessary to understand the origin of lunar features.

The main portion of the Atlas consists of 68 sections on individual regions on the side of the Moon visible from Earth. The sequence within the Atlas corresponds approximately to the way in which they may be observed during a lunation, beginning with the narrow crescent shortly after New Moon, through to Full Moon. This order follows sunrise on the Moon. After Full Moon, the formations vanish from sight in a similar sequence, until just a tiny crescent remains. An additional section is devoted to the farside of the Moon, invisible from Earth.

Keys to the location of all the principal formations described may be found on the front and rear endpapers.

A total of 388 images show maria, craters, mountains and rilles under various lighting and libration conditions, to give a three-dimensional picture of a particular region. We have tried to attain a good, high, and approximately consistent standard of quality in choosing the images. Naturally, however, the scale, resolution and degree of detail vary according to the instrumentation employed, as well as because of the conditions prevailing at the time of the exposure.

There are 422 lunar formations fully described in the Atlas sections. More than 100 additional features are mentioned in the text and marked on the photographs. The index at the end of this book lists, in alphabetical order, all of the features that are mentioned. In addition, the reader will find a glossary of the most important terms, as well as details of further reading and Internet sites.

Contents

Preface   p. iv

Advice on using this book   p. iv

The Moon – an introduction   p. 6

Atlas of lunar formations   p. 32

Glossary   p. 186

Index of lunar features   p. 188

Image credits   p. 190

Further reading and references p. 191

1	Mare Smythii	24	Montes Caucasus	47	Mare Nubium
2	Mare Crisium	25	Autolycus/Aristillus	48	Fra Mauro
3	Cleomedes	26	Cassini	49	Mare Cognitum
4	Endymion	27	Montes Alpes	50	Mare Insularum
5	Atlas/Hercules	28	Plato	51	Copernicus
6	Montes Taurus	29	Montes Teneriffe	52	Eratosthenes
7	Palus Somni	30	Archimedes	53	Mare Imbrium
8	Mare Fecunditatis	31	Montes Apenninus	54	Sinus Iridum
9	Langrenus/Petavius	32	Mare Vaporum	55	Gruithuisen
10	Mare Australe	33	Rima Ariadaeus	56	Mare Frigoris
11	Vlacq	34	Rima Hyginus	57	North Pole
12	Vallis Rheita	35	Sinus Medii	58	Aristarchus
13	Mare Nectaris	36	Hipparchus	59	Kepler
14	Rupes Altai	37	Ptolemaeus	60	Seleucus
15	Abulfeda	38	Rupes Recta	61	Reiner
16	Theophilus	39	Regiomontanus	62	Letronne/Hansteen
17	Sinus Asperitatis	40	Maurolycus	63	Gassendi
18	Statio Tranquillitatis	41	South Pole	64	Mare Humorum
19	Mare Tranquillitatis	42	Clavius	65	Schickard
20	Mare Serenitatis	43	Tycho	66	Sirsalis
21	Posidonius	44	Schiller	67	Grimaldi
22	Lacus Mortis	45	Palus Epidemiarum	68	Mare Orientale
23	Aristoteles/Eudoxus	46	Pitatus	69	Lunar Farside

Physical data on the Moon

Data	Moon	Earth	Moon/Earth ratio
Surface area	37.9×10^6 km^2	511×10^6 km^2	0.074
Mass	0.07349×10^{24} kg	5.9736×10^{24} kg	0.0123
Volume	2.1958×10^{10} km^3	108.321×10^{10} km^3	0.0203
Equatorial radius	1738.14 km	6378.1 km	0.2725
Polar radius	1735.97 km	6356.8 km	0.2731
Mean radius	1737.1 km	6371.0 km	0.2727
Flattening	0.0012	0.00335	0.36
Average density	3350 kg/m^3	5515 kg/m^3	0.607
Gravitational acceleration at the surface	1.62 m/s^2	9.78 m/s^2	0.166
Escape velocity	2.38 km/s	11.2 km/s	0.213
Mean Bond albedo	0.11	0.306	0.360
Mean visual albedo	0.12	0.367	0.330
Magnitude at a distance of 1AU	+0.21 mag	−3.86 mag	-
Solar constant (outside the atmosphere)	1367.6 W/m^2	1367.6 W/m^2	1.000
Maximum topographic altitude difference	16 km	20 km	0.800
Mean magnetic-field strength at the surface	$\sim < 2 \times 10^{-3}$ G	0.31 G	$\sim$ 0.0032

Orbital parameters (relative to Earth)

Semi-major axis	384 400 km
Mean distance at perigee	363 300 km
Mean distance at apogee	405 500 km
Orbital eccentricity	0.0549
Sidereal period	27.3217 d
Synodic period	29.53 d
Mean orbital velocity	1.023 km/s
Maximum orbital velocity	1.076 km/s
Minimum orbital velocity	0.964 km/s
Inclination of Moon's orbit to the Ecliptic	5.145°
Inclination of Moon's orbit to the Earth's equator	18.28° to 28.58°
Inclination of Moon's polar axis to its orbit	6.68°
Increase in distance between the mass centres of the Moon and Earth	3.8 cm/year

The lunar atmosphere

Night/Day temperature difference	>100 K to <400 K, roughly −160°C to +130°C, as low as −200°C at the poles
Total mass	c. 25 000 kg
Atmospheric pressure [surface at night]	3×10^{-15} bar
Number of atoms/molecules	2×10^5/cm^3
Surface pressure	$\sim < 1 \times 10^{-9}$ hPa
Estimated number of particles	Helium-4 (40 000/cm^3), Neon-20 (40 000/cm^3), Hydrogen (35 000/cm^3), Argon-40 (30 000/cm^3), Neon-22 (5000/cm^3), Argon-36 (2000/cm^3), Methane (1000/cm^3), Ammonia (1000/cm^3), Carbon dioxide (1000/cm^3) and traces of Oxygen (O$^+$), Aluminium (Al$^+$), Silicon (Si$^+$) and possibly Phosphorus (P$^+$), Sodium (Na$^+$) and Magnesium (Mg$^+$)

Structure

As with terrestrial seismology, conclusions may be drawn about the inner structure of the Moon from the distribution, strength and propagation rates of lunar seismic waves. Everything that we currently know about the inner structure of the Moon is based on the measurement of seismic waves, originating from Moonquakes, meteorite impacts, or artificially induced impacts by targeted crashes of lunar probes or carrier-rocket stages. The seismometers were installed on the Moon by the Apollo missions (Apollo 12, 14, 15 and 16), and were active until the end of 1977. Additional information has been provided by the evaluation of long-term Lunar Laser Ranging measurements.

Like the Earth, the structure of the Moon consists of a crust, a mantle beneath that and, at the centre, a small core. The lunar crust on the nearside, turned towards Earth, primarily consists of a feldspar anorthosite that is between 50 and 70 km thick. In the lunar highlands the uppermost layer, approximately 20 km thick, is the so-called 'megaregolith', a loosely structured bedrock layer, broken by cracks and fractures. When compared with the Earth, the Moon's crust is about three times as thick.

The crust exhibits one abnormality, in that it is, on average, almost double the thickness on the opposite side to Earth, which means it is between about 100 and 140 km thick. This significantly thicker lunar crust explains why the nearside and farside appear so different: there are practically no large lava flows on the farside. Most of the impact events were not powerful enough to create fractures and cracks that penetrated sufficiently deep into the thicker crust, to open feeder channels allowing lava from the mantle to rise to the surface.

There is, however, as yet no accepted scientific theory that explains the differing thickness of the lunar crust. Charles J. Byrne attributes the differences in the crustal thickness to the creation of a 'mega-basin', early in the history of the Moon's development. According to Byrne a gigantic impact on the nearside of the Moon, with estimated co-ordinates of 8.5° North and 22° East, created a basin with a diameter of over 3000 km and a depth of more than 20 km. Because of the enormous force of the impact, the impactor must have practically forced the lunar crust on the nearside down into the mantle over a good part of the surface. This theory is not undisputed, because according to model calculations such an enormous impact would have broken up the young Moon, and the Earth would probably now have a ring system like that of Saturn.

Between the lunar crust and the core, nowadays there is a solid mantle, consisting of olivine- and pyroxene-rich basalt material, with a higher density (c. 3.3–3.5 g/cm³) than the anorthosite of the crust (c. 1.0–3.0 g/cm³). Between the crust and the mantle there lies a thin layer of KREEP material, that exhibits an increased fraction of thorium and uranium. At a depth of about 500 km, seismic waves change velocity, which indicates a change in the density of the mantle material. This means that the mantle is also differentiated into an outer and an inner zone.

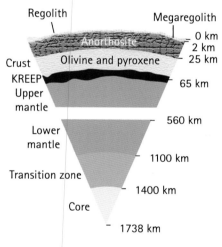

The internal structure of the Moon.

According to evaluation of lunar-laser-ranging results in recent decades, the Moon's core has a diameter of just under 700 km, is probably fluid, with a temperature of around 1600°C, and consists largely of iron and sulphur. The Moon's centre of gravity is displaced from the geometrical centre point by about 2 km in the direction of the Earth. When compared with the Earth's core, that of the Moon is very small. It amounts to only about 25 per cent of the Moon's radius, whereas the Earth's core has a radius that is 50 per cent of the Earth's radius.

Surface

Much of what was known at the end of the last century about the mineralogy, the geology and as a result, about the sequence over time of the formation of the lunar surface and its structures, was based on investigations of the few surface samples from the Apollo and Luna Moon missions. But these originated is just a few areas of the surface, and therefore provided only crude clues about the overall lunar surface. Samples from mare and highland areas were available.

The surface rocks from the highlands (anorthosite) are poor in iron oxide and titanium dioxide. The basalt mare lavas are rich in iron oxide and the fraction of titanium dioxide is variable.

Chemical composition of the lunar surface	
Element	Fraction
Oxygen	42%
Silicon	21%
Iron	12%
Calcium	8%
Aluminium	7%
Magnesium	6%
Titanium	3%
Others (together)	1%

Colour tints on the lunar surface enable conclusions to be drawn about the composition. In the visible region, the anorthosite on the lunar crust appears bluish, and the mantle lavas have varying reddish tints according to the fraction of titanium dioxide present. An exception is Mare Tranquillitatis, which appears blue, because of the high titanium–oxide fraction in the lava.

The basalts are thus subdivided into those rich in titanium dioxide (more than 7 per cent by volume); poor in titanium dioxide (2 to 7 per cent), and very poor in titanium dioxide (< 2 per cent).

Towards the end of the 1980s it was recognized that iron oxide, in particular, but also titanium dioxide – depending on its distribution in lunar rocks – altered the reflection of sunlight, dependent on wavelength. Basalt lavas with a higher titanium-dioxide concentration (e.g., Mare Tranquillitatis) reflected better in the blue spectral region. The albedo of titanium-poor basalts increased at the red end of the spectrum. A similar effect occurred with the albedo of highland rocks. The titanium-dioxide concentration in mare basalts is thus an indicator that the lava flows occurred at various lunar epochs and originated at different depths within the Moon's interior. The reflectivity of other elements or their oxides also alters as a function of wavelength (not just in visible light). This showed, for example, that the basalt lava in the South Pole-Aitken Basin has a significantly lower iron content than the basalt lavas on the nearside.

At the beginning of the 1990s, planetologists at the University of Hawaii, among others, developed methods of determining the iron and titanium concentration of surface rocks from monochromatic lunar photographs. These monochromatic images were subsequently colour-coded and combined, and thus immediately revealed visually the distribution of iron and titanium concentrations. Wavelengths of 750 nm and 950 nm (near infrared) were used for iron oxide, and 415 nm and 750 nm for titanium dioxide. This procedure was later calibrated with soil samples from the Apollo landing sites.

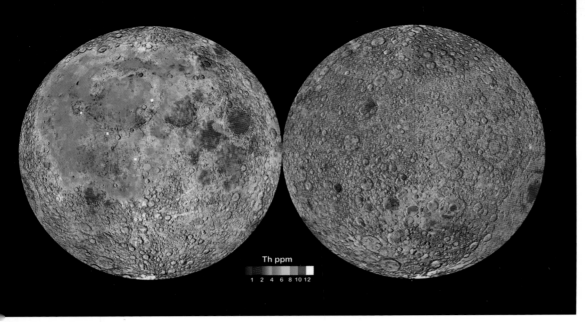

The distribution of thorium across the surface, shown colour-coded for both hemispheres.

Lunar rocks

Until 2010, the analysis of soil samples from the Moon showed no signs of water molecules. In addition, lunar rocks were distinguished from terrestrial minerals by an extremely low fracton of volatile elements. These two facts were the principal factors in favour of the theory that the Moon arose from a grazing impact of a body with the early Earth, and that, shortly after its formation, it was completely or at least extensively molten.

In May 2011, to some surprise, researchers from Brown University (Rhode Island, USA) announced, in the journal *Science Express*, the discovery of water in lunar rock samples, in particular in the titanium-rich 'orange soil' glass spherules that were collected by the geologist Harrison Schmidt from the crater Ching-Te during the Apollo 17 mission – and in quantities that were even higher than in volcanic rocks from the Earth's lower mantle. If these measurements are confirmed they will provoke a renewed discussion about the origin of the Moon. Or might the evidence of water be reconciled with the impact theory?

The results of this chemical spectroscopy were later supplemented by the Galileo, Clementine and other Moon missions. Clementine provided a global picture of the Moon, obtained in eleven spectral regions with an average resolution of 200 m/pixel. Similarly, the Clementine mapping of the gamma-ray spectrum revealed a significantly enhanced thorium concentration in the basalt lavas of Oceanus Procellarum and Mare Imbrium. Thorium is a radioactive element and a component of the so-called KREEP material. A thin layer of KREEP-containing rock is thought to lie beneath the lunar crust, and the deposits on the surface imply that the impacts that created Oceanus Procellarum and Mare Imbrium, must have penetrated the lunar crust to a depth of at least 50 km.

The regolith and megaregolith

The whole upper surface of the Moon is covered with regolith. The term 'regolith' is not a description of a particular mineral or rock, but rather a term from geomorphology, and simply means a loose layer that has formed from the underlying layers of rock.

The Moon's regolith has mainly arisen from the anorthositic rocks of the Moon's surface, created over hundreds of millions of years by the continuous bombardment by everything from large impacts down to the permanent rain of micrometeorites. The lunar regolith has a consistency that is comparable with that of flour or icing sugar and a density of c. 1 g/cm³ to 3 g/cm³ at the surface. The highland regolith is intermingled with aluminium-rich anorthosite (a silicate/feldspar rock). In colour images the material is bluish with a very high albedo.

In the oldest lunar regions in the southern highlands, the regolith layer may (at its greatest) reach a depth of about 2 km. As a result this layer is known as the megaregolith. Lunar samples, brought back to Earth by Apollo 16 (from the highlands in the Descartes region), have been dated to an age of about 4.2 × 10⁹ years. To a depth of a few metres the surface is like powder, and below that consists of rubble of ever-larger blocks.

Because the cratering rate had already sharply decreased by the time that the lava flows occurred in the major impact basins, about 3.5 × 10⁹ years ago, the regolith layer over the younger lava flows is only about 1 to 2 m thick. Statistical studies have established that, in highland areas, the younger craters are less numerous and also have smaller central peaks than older craters in these regions. An explanatory model suggests that the immense pressure at the centre of the impact (which eventually leads to the formation of central peaks), is dispersed sideways and weakened in the 2 km thick, loose structure of the megaregolith, which has, of course, become even thicker over the course of time. Naturally, only smaller craters have resulted from the lesser impact energy, although Tycho – one of the youngest lunar craters – has a magnificent central peak.

In total 382 kilograms of lunar rocks were brought to Earth by the Apollo missions, and most of these were obtained by the last two flights: Apollo 16 and Apollo 17. The three Luna probes launched by the Soviet Union: Luna 16 (eastern portion of Mare Fecunditatis), Luna 20 (highland between Mare Crisium and Mare Fecunditatis), and Luna 24 (Mare Crisium) together brought just 326 g of lunar rocks back to Earth. But because these samples originated in different regions from those of the Apollo missions, they are, of course, just as important. To these may be added the few fragments that undoubtedly come from the Moon, and which have been found on Earth – primarily in Antarctica.

The surface samples collected by the Apollo missions may, in general, be divided into two different categories: basalts from the low-lying mare surfaces, and the significantly brighter anorthosite from the lunar highlands. Age determinations show that the basalts are about 3.2 to 3.8 × 10⁹ years old, and are thus considerably younger than the anorthosite, which may be as old as 4.5 × 10⁹ years.

Anorthositic lunar rocks consist of a high proportion by volume of anorthosite itself, a calcium-rich plagioclase feldspar (rather rare on Earth), and is thus frequently described by geologists as a plagioclase feldspar rather than an anorthosite. Other components of this anorthosite may be pyroxene, olivine, and metallic ores, such as magnetite, in various concentrations. The term 'basalt' normally means nothing more than a dark rock is involved, one that has solidified from fluid lava. The term 'mare basalt' implies that the type of basalt is primarily found on the surface of the lunar maria. If, for example, a basalt shows a high content of radioactive KREEP elements, then geologists speak of a KREEP-basalt. This basalt was brought to the surface from the deep layer through impact events, and is primarily located on the surface of mare areas, with high concentrations in Oceanus Procellarum and Mare Im-

Probably the most sensational rock returned from the Moon by Apollo 15, is the so-called Genesis Rock, sample number 15415. It weighs 270 g, consists of anorthosite and dating gives an age of 4.5 billion years. It comes from the foot of the Apennine mountain range. It is the only lunar rock to have such a great age.

A typical example of mare basalt, the dark colour of which causes the maria to appear dark. The sample is vesicular, which indicates that the fluid lava contained gas under high pressure. When the lava flow reached the atmosphereless surface, the pressure would have decreased, and the gas expanded to form the bubbles.

The significantly lighter anorthosite from the lunar highlands provides a contrast. It consists primarily of feldspar, a mineral that itself consists mainly of silicon, calcium and aluminium. It is normally less dense than the iron-rich mare lavas.

brium. Anorthosite was the first to crystallize shortly after it originated in the widespread melting of the Moon. Because of its lower density, it rose to the surface of the lava and formed the Moon's crust. In the maria, the anorthosite lies beneath the lava layers. It forms the oldest lunar rocks.

About 10 per cent of the old highland regions consist of minerals rich in magnesium. These are the so-called plutonic rocks (primarily granite-like), which are normally produced at very great depths (even below the former mantle lava). If cooling of the lava takes place beneath the surface, they are described as being intrusive. Because of the relatively high insulation provided by the overlying rocks, the magma cools only slowly, producing large crystals of the various minerals. They exhibit an admixture of the rare earths and are probably related to these. They were probably brought close to the surface by heat and convection, but how they came to be in the lunar crust is, so far, uncertain.

Origin and evolution

Any theory of the origin of the Moon will never be susceptible to exact scientific proof, but any theory that may be tested by theoretical simulations must take the following facts (among others) into account:

- Whereas all the other satellites in the Solar System amount to just a few hundredths of a percent of their parent planets, the Moon, in comparison with the Earth, is relatively enormous with a mass of about 1 per cent of the Earth's mass.
- The Earth's mean density is 5.5 g/cm^3, but that of the Moon is only 3.3 g/cm^3. The current interpretation of this difference is that the Earth has a massive iron core, whereas the Moon, by contrast does not have an iron core, or has just a very small one.
- The overall angular momentum, resulting from the Earth's rotation and the Moon's orbit around the Earth is, in comparison with the other planets and their satellites, extremely large.
- When compared with the Earth and meteorites, lunar rocks have a significant deficiency in calcium and sodium.
- The isotopic ratios of various oxygen molecules in terrestrial and lunar rocks is absolutely identical, but that of the Earth and Moon differs from various meteorites. It follows that the Moon and the Earth arose from very similar material.

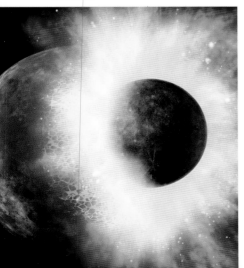

The most popular theory of the Moon's origin proposes the impact of a body ('Theia') with the Earth. The Moon originated from the ejected remnants.

- The upper 200 km of the body of the Moon, and possibly even the whole Moon, must have been completely molten shortly after its origin. All lunar rocks originated in high-temperature processes.

At present, no serious planetologist doubts the following account of the formation of the Moon: About 4.5 billion years ago, the early Earth, which had already differentiated into crust, mantle and core, experienced a grazing collision with an asteroid-like body of about 5000 km in diameter (about the size of Mars). This asteroid is often given the name Theia in the specialist literature. (Theia was the Greek goddess of light and her daughter was the Moon goddess Selene.) In this grazing impact Theia was partially vaporized, partially melted and was literally completely pulverized. A portion of the material fell onto the Earth's surface, Theia's heavy core sank into the Earth's core, and the remaining material gathered into a disk around the Earth. Model calculations indicate a diameter of the disk as between 25 000 and 30 000 km. From the material in the disk condensed – in accretion driven by gravity – and part of Theia became the Earth's Moon. Over the course of time the surface of the Moon cooled and the lunar crust formed over the still-fluid mantle. The first minerals crystallized out; and the lighter ones rose into the crust and the denser ones sank into the mantle.

A phase known as the Late Heavy Bombardment followed, and which has been dated to a period between 4.1 and 3.8 billion years ago. During this period, all the inner planets, including the Earth and the Moon, suffered heavy bombardment by residual material (planetesimals with diameters of between 1 km and

Periods in the history of the Moon				
Period	Age in billions of years	Duration in million years	Event(s)	Examples
Pre-Nectarian	4.5 – 3.92	780	The time between the origin of the Moon with the subsequent formation of the crust and mantle, to the formation of the Nectaris basin.	Mare Tranquillitatis, Ptolemaeus, giant impact basins, Grimaldi
Late Heavy Bombardment	4.1 – 3.8	300	Formation of many of the lunar craters and basins.	
Nectarian	3.92 – 3.85	70	The transition to the Imbrium period was marked by the formation of the Imbrium basin.	Mare Crisium, Mare Humorum, Mare Nectaris, Clavius
Early Imbrium	3.85 – 3.80	50	Toward the end of this era, the Orientale basin was formed. It is thought to be the youngest of the large lunar basins.	Petavius, Cassini, Macrobius, Arzachel
Late Imbrium	3.8 – 3.15	650	The period during which the major lava flows filled the impact basins. Towards the end of the Imbrium epoch, most of the areas of Dark Mantle Deposits were created by the eruption of pyroclastic lavas.	Archimedes, Plato, Posidonius, Piccolomini, Sinus Iridum
Eratosthenian	3.15 – 1.0	2150	At the beginning of this epoch, the major lava flows slowly ceased, and there was a decrease in the high crater-forming impact rate. Craters such as Eratosthenes were formed in the solidified lava flows.	Eratosthenes, Pythagoras, Bullialdus, Theophilus, Langrenus
Copernican	1.0 – today	1000	The ray systems of craters such as Eratosthenes were almost completely darkened through erosion. Craters with ray systems that are still bright were created in this period. Copernicus has been dated to an age of c. 800 million years, and Tycho to c. 100 million years.	Copernicus, Tycho, Aristarchus, Kepler, Eudoxus

In the northern part of Mare Nectaris there is an area where all five classes on the Arthur scale are found close together. Craters such as Theophilus or Mädler are believed to be young. They appear bright under high solar illumination. Other criteria are a ray system or a halo of bright ejecta and crater wall without breaks and with sharp slopes.

The Arthur Scale			
Class	Description	Examples	Age in billion years
1	Fresh rims, rays	Theophilus, Mädler, Kant, Alfraganus	0 to 2.9
2	Freshest post-Mare	Bohnenberger	3.0 to 3.4
3	Softened rims	Capella, Isidorus, Ibn Rushd, Cyrillus	3.5 to 3.7
4	Heavily degraded	Torricelli R, Daguerre	3.8 to 4.0
5	Faint outline	Structure beneath Torricelli C	4.1 to 4.5

The principle of lunar superpositioning is extremely simple. It is often possible – without an exact knowledge of the geological age of structures – to recognize the rough sequence of a surface's creation. The corresponding region is examined in terms of the Arthur scale and the surface features arranged into a time sequence. In this, the following criteria are important:
- the crater density (number of craters per unit surface) in the corresponding region,
- the appearance (brightness) of the ray systems of craters,
- the deposition of ejecta and small craters on the crater walls of larger craters, and
- damage to craters though later impacts.

The crater density in older areas of the lunar surface is significantly higher than in younger areas; craters with bright ray systems or a bright halo are younger than craters without ray systems or haloes. Small craters on crater walls are younger than the crater walls and other deposits on crater walls (debris) must be younger than the original crater. The method of superposition is naturally susceptible to widespread, later alterations to the landscape, but, in general, it is very effective.

50 m) and the numerous lunar craters were formed. The largest of these bodies excavated the major basins in the lunar crust, which were later filled with lavas from the mantle. Before the major lava flows the crater-formation rate had declined steeply, and since about 3 billion years ago there have been only occasional eruptions of lava and active volcanism.

Determining the ages of craters

In the 1960s and 70s, when there was an increase in scientific lunar research in preparation for the Apollo missions, David Arthur of the University of Arizona was working on crater catalogues and statistics. He modified the old classification by Baldwin and introduced the 'Arthur Scale' to categorize surface features into various lunar epochs. Crater ages were classified in 5 stages from very young to highly eroded.

Over the course of millions of years, craters and their crater walls are slowly destroyed and flattened through lunar erosion. The formerly bright ejecta that forms a halo or ray system darkens, until it can hardly be distinguished from the surrounding landscape. Class 5 corresponds to structures of which only ruined portions remain. They are generally recognizable only under grazing illumination. Many former craters practically no longer exist because they have been destroyed.

Stratigraphy and the superposition method

In the attempt to develop a lunar stratigraphy, and thus dating of the age of rocks and the succession of geological eras, what was missing were comprehensive lunar rocks from the widest range of regions of the Moon's surface. In 1962, Shoemaker and Hackman in the USA realised, however, that impact events and their ejecta served as perfect time markers for a lunar stratigraphy. It is possible to differentiate between old and young and thus between earlier and later events. This introduced the principle, known as 'superposition' or 'superpositioning', and which in lunar research is partly based on the Arthur scale, and which, after 1969, could be calibrated through age determinations of lunar surface samples.

The region around the very young crater Copernicus and the older (on the Arthur scale) crater Eratosthenes was chosen by Shoemaker and Hackman for drawing up the first geological map using the superposition method. This region was a suitable choice because of the numerous lava flows in the Mare Imbrium, Sinus Aestuum, and the many secondary craters in the surrounding area.

Over a period, in conjunction with the numerous results from lunar probes and the analysis of rock samples from the Apollo, Luna and Lunakhod missions, up to the end of the 20th century, 44 geological stratigraphy maps had been produced using the superpositioning principle. Through the multispectral mapping of the Moon's surface by the Clementine lunar probe, the stratigraphy of the whole of the lunar surface has now been recorded.

The formation of craters

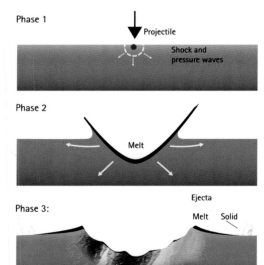

The phases of the Standard Sequence for the formation of lunar craters. After Wood, 2007.

Among experts it is now undisputed that nearly all lunar craters (well over 99 per cent) as well as the lunar basins were created by the impact of asteroid-like bodies, and that the overall, basic structure of a crater is primarily only dependent on its diameter (and, with it, the diameter of the impacting body).

In the 1960s, the so-called Standard Sequence for lunar craters was defined by the lunar geologists and planetologists of the USGS (United States Geological Survey). It fundamentally consists of three elements:

Phase 1 – Contact/compression stage

The impacting body hits the solid surface of the outer crust, and compresses it vertically and also horizontally. High-energy shockwaves propagate through both the impactor and the crust, and vaporize, melt, shatter, and pulverize the lunar surface as well as the impactor. The pressures thus induced are dependent on the impact velocity and the consistency and mineral composition of the crust at the point of impact. They amount to between one hundred and a few thousand gigapascals. The temperatures that melt the impacting body and the rocks at the point of impact, and vaporize those at the centre, amount to a few tens of thousands of degrees Celsius. The duration of the compression phase depends on the velocity and size of the impacting body and, on average, lies between one thousandth and a few tenths of a second.

At the end of Phase 1, the inner surface of the (stationary) point of impact has absorbed the immense kinetic energy of the impact and converted it to heat and pressure, and the impacting body and crust have both fused. The actual crater is first formed in Phase 2.

Phase 2 – Excavation stage

In this phase the hemispherical shockwaves propagate deep into the interior of the Moon, compressing the rocks even more. The shockwaves are reflected by deeper layers and race back to the origin of the impact. When the shockwaves reach the surface, they dissipate because they are then able to expand unimpeded. This decompression of the shock waves corresponds to a gigantic explosion, the crater is formed, and the ejecta, consisting of debris, boulders, and molten surface rocks are flung out radially and fall back onto the nearby surrounding area of the violently quaking surface. This material forms an ejecta blanket around the crater that has been formed. With the release of pressure, the compressed lunar surface 'bounces' back and the crater wall is raised. About a quarter to one half of its height may be attributed to this upward bulge, and the remainder of the final height of the crater wall consists of the rocks flung out of the interior of the crater, and where the uppermost layer of the crater wall consists of the deepest rocks from the crater.

This second phase lasts a few minutes, during which the diameter and depth of the crater increase. At the end of Phase 2, the crater that has been formed through decompression has become about a magnitude larger in diameter than the original diameter of the impacting body (between 1:10 and 1:20). The end of Phase 2 is also known as the Transient stage.

Phase 3 – Modification stage

In simple craters up to diameters of about 15 km, hardly any significant changes occur in the crater. Gravity may perhaps cause loose rubble from the crater wall to fall down onto the crater floor, filling it up and flattening the floor. But such changes may also occur much later – triggered by moonquakes or the shock waves from later impacts. In addition, lava flows that occur later may modify the crater. This third phase lasts no more than about 10 minutes with simple craters.

Changes by erosion

Frequency of current crater formation	
Crater diameter	Number per m²/year
1/10 000 mm	~ 300 000
> 1/1000 mm	~ 12 000
> 1/100 mm	~ 3000
> 1/10 mm	~ 1
> 1 mm	~ 0.001

Small-scale structures on the lunar surface are permanently altered by four primary factors:

- the permanent 'rain' of micrometeorites,
- the solar wind and cosmic rays,
- the great temperature changes, and
- moonquakes

Although the effects of micrometeorites may be minimal over short periods of time, over long stretches of time, of hundreds of millions of years, they have caused initially sheer crater walls and central peaks have now become rounded and flattened.

The regular temperature change from c. +130°C and -150°C between lunar day and night also plays a part in the degradation of the uppermost layer of the regolith on the Moon's surface, by pulverizing the rock. At a depth of 1 m below the surface constant temperatures of about -35°C prevail. At present the surface of the Moon is covered in a layer of fine, dispersed anorthosite, so-called moondust, many metres thick.

Finally, moonquakes, triggered by solar and terrestrial tidal forces that affect the interior of the Moon, may also cause small-scale changes, in that loose rubble from the inner crater walls collapses onto the crater floors. Because these quakes are weak, they do not cause widespread changes.

The small crater Linné was the subject of controversy over changes in the lunar landscape.

Lunar Transient Phenomena

Lunar transient phenomena (LTP) are also known in the literature as TLP. These are fleeting lunar phenomena and involve observations of short-lived changes on the lunar surface. The history of LTPs probably began with a report by the experienced lunar observer, Julius Schmidt, in 1866, when he stated that the crater Linné, lying in the western central region of Mare Serenitatis, had disappeared. The then current Moon map by Beer and Mädler from 1837 showed Linné as a crater c. 10 km across, surrounded by a small, but very bright halo of ejected material. The Moon map by Lohrmann from 1824 also showed the crater. Linné is a difficult object for small telescopes, because the halo consist of bright material.

Schmidt's publication caused a sensation. Astronomers, who had believed that with the publication of Beer and Mädler's map lunar research had come to an end, turned their telescopes back to the Moon. A flood of observations followed, which culminated in the report that now, at the location of the crater, a small mountain peak had arisen in the middle of the halo. Given the date at which these reports originated, it must be borne in mind that lunar observers expected such changes, similar to those that occurred on Earth through geological processes. These observational descriptions culminated in those by von Gruithuisen, who frequently described enormous, artificial structures, constructed by the inhabitants of the Moon.

The reports by Schmidt and many others about the changes in the crater Linné may certainly be attributed to inadequate telescope optics and poor seeing conditions. In the following decades, there were a few such observations of mysterious changes. Many of these involved the interior of the crater Plato; observers maintained having seen obscuration of the crater floor by superficial dust clouds. Others reported that the floor of the crater was covered with myriads of glittering points of light (like sunlight reflected from a water surface).

At the beginning of the 20th century, reports accumulated from lunar observers who claimed they had seen colour changes or luminous patches (or both) that a short time later had become invisible. Because it was then thought that lunar craters were of volcanic origin, these observations were ascribed to active volcanism. Such observational reports were, until the end of the 1950s, simply ignored by the few lunar geologists. Observation of the Moon had no part in professional astronomy, and it was left to the amateurs. At that time, only a few professional astronomers confined themselves to lunar research.

The situation first changed with the observations by D. Alter, an astronomer at Mount Wilson Observatory, and the Russian astronomer N. A. Kozyrev. In 1956, Alter took a series of photographic images of the craters Alphonsus and Arzachel in various spectral regions with the 60″ telescope. He thought that he could see distinct differences in sharpness between the rille systems in the two craters in images taken in the blue spectral region. Those in Alphonsus seemed distinctly less sharp that the rilles in Arzachel. Because the small craters along the Alphonsus rilles were undoubtedly of volcanic origin and were not impact craters, he believed that the blurring could be attributed to emission of gas from the craters.

Subsequently, Kozyrev began regular spectroscopic of this region of the Moon in 1958. In some spectra taken on 3 November 1958, he found absorption lines of gaseous carbon. Kozyrev announced these spectra as showing a lunar eruption, and Alter pronounced that this observation and the photographs of the Moon's farside taken by Luna 3 as the most important that had ever been made.

When at the beginning of the 1960s it became clear that men would fly to the Moon, such events were investigated in detail. Most of the reports of LTP related to regions that could be definitely linked to former volcanism, such as the Aristarchus Plateau or the interior of Alphonsus. And the events frequently occurred at times when the gravitational forces of the Sun and the Earth reinforced one another, and thus affected the lunar surface particularly strongly (just as the seismometers left by the Apollo missions have recorded stronger moonquakes at such times).

In 1963, increasingly significant observations followed. J. Greenacre and E. Barr carried out lunar observations with the 24" telescope at the Lowell Observatory to draw up maps in preparation for the Apollo Moon landings. On 30 October, for 25 minutes they observed the appearance on the Aristarchus Plateau of three orange-red flickering points of light on the southwestern crater wall of Aristarchus, in the immediate neighbourhood of the crater where Vallis Schröteri begins, and also at the first bend of Vallis Schröteri. Exactly a month later, Greenacre and Ball, together with three other observers simultaneously using another telescope, observed luminous events in the same area, this time reddish-purple colours, lasting for 75 minutes. Both observations were, at the time, interpreted as volcanic outbursts, but nowadays this is thought to be rather dubious.

Lunar geologists now think that at least some of the LTP reports should be considered realistic – but that they should not be attributed, however, to active volcanism, but to the escape of gases created by the radioactive decay of uranium and thorium. In any case, none of the many hundreds of thousands of images that have been taken by lunar probes up to now, show any signs of active volcanism. During their lunar orbits, the crew of Apollo 15 measured active escapes of gas. When flying over the Aristarchus Plateau a high concentration of radon gas was measured. High values were also discovered over the Ina region.

Nomenclature

The nomenclature that is still, in principle, used today was introduced by Giovanni Battista Riccioli (1598–1675), an Italian theologian, philosopher and astronomer. In 1651 Riccioli published his *Almagestum Novum*, and in it a map of the Moon (albeit based on the work by his pupil Francesco Grimaldi), in which he introduced the basis for lunar nomenclature.

Following the precedent of a Moon map by the Danzig astronomer Johannes Hevelius (1611–1687), Riccioli designated the dark areas as 'maria' (sing. 'mare') and gave them names of emotional states, such as Mare Crisium, 'Sea of Crises', or Mare Serenitatis, 'Sea of Serenity', or else descriptions of terrestrial weather – for which the Moon was then regarded as being responsible – such as Mare Frigoris, 'Sea of Cold', or Mare Imbrium, 'Sea of Rains'. He described bright regions as 'terrae'.

He named mountains and mountain ranges after terrestrial models, and craters – after the precedent set by a Moon map by Langrenus (Michael Florent van Langren, 1598–1675) – after famous personalities, such as astronomers, theo-

Dark lava plains (maria) and bright highlands (terrae) are the basic features governing lunar nomenclature.

logians, philosophers, etc. In doing so, he arranged the persons' dates in a spatial sequence – from north to south. These bases for lunar nomenclature have been retained to this day.

Significant expansions, driven by the introduction of better telescopes and as a consequence, better visibility of smaller surface features, were made by the lunar observers (the selenographers) Johann Hieronymus Schröter (1745–1816) in 1791 and 1802, and subsequently by Wilhelm Beer (1797–1850) and Johann Heinrich Mädler (1794–1874) in their *Mappa Selenographica* (published in 1837). Beer and Mädler's Moon map had the names of 437 craters, mountains and maria. Of these 200 descriptions were taken from Riccioli and 60 from Schröter. A further 145 names, including those of famous geographers and sailors, were added by Beer and Mädler.

Beer and Mädler extended lunar nomenclature, in that they designated smaller craters near (or within) large craters by Latin capital letters (such as Clavius C, D, J, K, L, and N). Other structures, such as hills, for example, in the neighbourhood of large craters, were designated with lower-case Greek letters. Such, for example, are the two central peaks in the crater Langrenus, which were designated α (alpha) and β (beta). Further examples are the lunar domes Arago α and β, or Kiess π (pi). Rilles in rille systems were designated with Roman numerals. Normally craters carry just the corresponding person's surname. Where the possibility of confusion existed, abbreviations of the first names were introduced. An example is the Herschel family, which is represented on the Moon by three prominent craters: Herschel (Friedrich Wilhelm), his sister Caroline (C. Herschel) and his son J. Herschel.

Later lunar observers admittedly adopted this nomenclature, but also arbitrarily changed or altered names – sometimes out of self-interest as well. To resolve the confusion over different names on various Moon maps, the International Astronomical Union (IAU), which was founded in Paris in 1919, gave the astronomers M.A. Blagg and K. Müller the task of drawing up a list with unified and internationally binding names for lunar surface features. It contained 681 names on the nearside of the Moon. Since then, the IAU is the sole institution that is internationally authorized to award names for the surface features of the Moon (and the planets), or to alter names. The development of global lunar mapping (including the farside of the Moon) necessitated an extensive expansion of the list of names. In 1970, at the 14th Congress of the IAU in Brighton, what was then Commission 17 of the IAU adopted 513 new names. It was decided that only the names of deceased persons would be chosen who deserved recognition for having made extraordinary scientific advances or discoveries. One of the few exceptions to this rule was the naming of twelve small craters after people who were then still living: six American astronauts, and six Soviet cosmonauts were thus honoured. So, for example, as a memorial and in recognition of the first manned

Lunar craters are named after deceased persons. Some of the few exceptions are the three craters Armstrong, Aldrin and Collin near the Apollo 11 landing site.

The designation of lunar features		
Latin designation	Translation	Example(s)
Catena	Crater chain	Catena Davy, Catena Abulfeda
Dorsum, Dorsa	Mare wrinke ridge (pl. Dorsa, used to indicate a system of multiple wrinkle ridges)	Dorsa Smirnov (Mare Serenitatis) Dorsum Oppel (Mare Crisium)
Lacus	Lake	Lacus Mortis
Mons, Montes	Mountain, Mountains	Mons Pico, Montes Apenninus
Oceanus	Ocean	Oceanus Procellarum
Palus	Swamp	Palus Epidemiarum
Promotorium	Promontory (Cape)	Promotorium Heraclides
Rima, Rimae	Rille, Rille system	Rima Hyginus, Rimae Triesnecker
Rupes	Slope, Escarpment	Rupes Altai
Sinus	Bay	Sinus Iridum
Vallis	Valley	Vallis Alpes

lunar landing by Apollo 11, small craters in Mare Tranquillitatis were named after Armstrong, Aldrin and Collins. The landing site was officially given the name of Statio Tranquillitatis (Tranquillity Base.)

A further, far-reaching change to lunar nomenclature was approved in 1973 at the IAU General Assembly in Sydney. The designation of subsidiary craters and other features in the neighbourhood of larger craters with Latin capital letters and Greek lower-case letters, introduced by Beer and Mädler was abolished. The reason for this was the work carried out in drawing up a detailed Moon map by NASA at a scale of 1:250 000, the Lunar Topographic Orthophotomap (LTO). The individual sheets of this map, which depicted only very small sections of the lunar surface, needed to show at least one feature with a specific name.

In total 138 craters, which formerly had been designated with letters, were newly named. So, for example the crater Olbers A became the crater Glushko, and Manilius A was given the name Bowen. For the first time international female and male first names were also adopted (such as Vallis Christel and Vallis Krishna in Mare Serenitatis). As early as 1976, the IAU partially rescinded this decision. On Moon maps that showed the new names, the former designations with Latin capital letters and Greek lower-case letters were reinstated in parentheses – to retain continuity with older literature and Moon maps. By February 2010, the IAU had officially named 8986 surface features.

Further identification of surface structures

Alongside the naming of the main features on the lunar surface – the maria, mountains and craters – some features were grouped together under Latin names. Some of these carried individual designations (e.g., mare wrinkle ridges), but many other had names that were related to nearby larger and more prominent features.

Frequently, the classification of individual lunar features into uniform groups of geologically similar formations was not followed consistently. For example, Vallis Rheita and Vallis Alpes were included in the same class – lunar valleys – whereas they are completely different formations with respect to their true geological structure (and origin). Vallis Alpes is the result of floor collapse between two linear fault-zones, whereas Vallis Rheita is a heavily eroded crater chain, which consists of partially overlapping craters, and must, therefore really be called Catena Rheita.

Topography

Even when observing the Moon with the naked eye it is obvious that the surface is divided into bright and dark regions. Early lunar observers (e.g., Galileo) thought that in the dark areas they were seeing oceans and seas, and in the bright ones, land surfaces (terrae). Later, when telescopes had improved, small craters and other structures were recognized in the monochrome, dark patches, and it was concluded that these areas could not be dried-out seas. It is now known that the maria are giant areas of basalt lavas, that may, in part, be as much as 3 km thick, and have been built up from many layers.

Mare/Maria

The lava flows are significantly younger than the Moon's highlands (terrae). Their formation is dated to a time about 4 to 3.8 billion years ago. From a geological point of view they are the central area of considerably large structures, the impact basins.

The basins were created by the impact of large, asteroid-like bodies early in the history of our Solar System. These impacting bodies were so large (possibly between 50 and 100 km), and the energy in the impact was so great that the crust of the Moon was penetrated. Many hundreds of years later,

Mare Nectaris is a typical multi-ring basin with several crater walls.

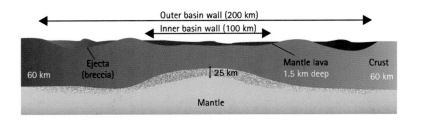

Outer basin wall (200 km)
Inner basin wall (100 km)

Ejecta (breccia) — 60 km — 25 km — Mantle lava 1.5 km deep — Crust 60 km

Mantle

The section across the Grimaldi Basin clearly shows the structure of a lunar impact basin. After Wood, 2007.

when, through natural radioactive decay and the resulting warming, a portion of the Moon's upper mantle melted, the lava filled the central regions of the large impact basins.

These lava flows lasted – with occasional interruptions – a few hundred million years, and ended about 3.0 billion years ago. Since then – apart from occasional impacts, which led to the formation of normal craters – the evolutionary processes affecting the formation of the surface has come to a halt.

The maria lava flows cover a total area of about 30 per cent of the nearside of the Moon, but on the farside, by contrast, only about 2.5 per cent. The difference is possibly to be attributed to the fact that the lunar crust on the farside is, at about 140 km, significantly thicker than on the nearside (c. 50–70 km), which clearly hindered the emission of lava from the lunar mantle. A scientifically sound explanation for this difference in thickness is still lacking.

The seismic pressure and shock waves created by the strike travelled through the molten lunar crust and – depending on the energy of the impact – raised one or more circular, concentric crater walls around the centre of the impact. An analogy for the formation of these crater walls may be seen in the waves that are created by a stone thrown into mud or any viscous fluid.

Most of the lunar basins are the so-called multi-ring basins, with two or more crater walls, and where the transition between the largest craters and the smallest basins is rather blurred. On the nearside of the Moon, only the remnants of the concentric crater walls of multi-ring basins are recognizable, such as around Mare Nectaris and Mare Crisium. The complete form of such an impact basin with concentric crater walls and a lava-flooded central region is most clearly seen in Mare Orientale, of which, unfortunately, only a very small section of the Montes Cordillera and Montes Rook crater walls and a few small lava flows are visible on the nearside.

The lava-flooded areas in the basins are slightly convex and follow the curvature of the Moon's globe, and the lava flows are significantly thicker in the centre than at the edges. The upper surfaces are extraordinarily flat, with the maximum difference in height being, on average, less than 100 m. The thickness of the lava layers differs in individual maria and may amount to 3 km.

The diameter of the lava-flooded central regions of the impact basins lie between 250 kilometres (small mare, for example Grimaldi) and about 500–700 km (Mare Tranquillitatis, Mare Serenitatis). The diameter of the outer crater walls are significantly greater, in the case of Grimaldi about 400 km, and with Mare Orientale as much as 900 km. The largest impact basin on the Moon is the South Pole-Aitken Basin, which lies on the farside. Individual mountain peaks of the outer crater wall may be observed near the South Pole under favourable libration conditions. Measurements of the diameter vary – according to the source – between 2300 km and 2500 km, with maximum height differences of up to 16 km. The diameter is difficult to determine, because there is no pronounced crater wall. The South Pole-Aitken Basin is very old and the crater walls have been destroyed and overlain by subsequent large and small impact events.

Towards the end of the 1960s, by monitoring the flight paths of the Lunar Orbiter probes (i.e., determining their orbital parameters), variations from the average lunar gravitational field were determined over all the maria and also over large craters and the larger complex craters. The centres were named mascons (mass concentrations). Over these areas, gravity is stronger and the more powerful gravitational attraction affected the orbital paths of the lunar probes. From these changes in the flight paths it was possible to determine that the density of the lunar rocks in these areas was about 3.3–3.5 g/cm³. The mascons were caused by the enormous layers of the denser mantle lava in the basins and the relatively dense upward bulges in the mantle beneath the crust. By contrast, the surrounding rocks on the lunar crust (with the average gravitational field) have a density of only about 1.0–3.0 g/cm³.

Regions that had a lesser density than their surroundings were designated negative mascons. They are probably to be attributed to rocks of lower density below the lunar crust. Around the borders of many maria there a concentration of the so-called FFC craters (Floor Fractured Craters), and very extensive, linear rille systems are observed.

The Imbrium Sculpture

The concept of the Imbrium Sculpture was introduced in a publication in 1893 by Grove Karl Gilbert (1843–1918), a pioneer of lunar geology. Gilbert was the first to notice that many structures throughout the southeastern region of the Moon – chains of craters, valley-like formations, mountain ridges, and destroyed crater walls – were oriented radially to the centre of Mare Imbrium.

He attributed all this destruction and these changes to already existing structures to a single gigantic impact, in which the Imbrium Basin was created, about 3.8 billion years ago. Imbrium is one of the youngest lunar basins. He described the whole of this more-or-less simultaneous destruction through solid, viscous and molten ejecta from the Imbrium impact as the Imbrium Sculpture.

The effects of these secondary Imbrium impacts may be clearly seen in the from of valley-like formations on the southeastern wall of Alphonsus where it borders on Arzachel; in the crater Glydén, north of Ptolemaeus, and on the crater walls of Hipparchus, almost 1000 kilometres from the centre of Mare Imbrium. Particularly plain to see are the furrows, scratches, and rilles, in the region north of Rima Ariadaeus and north and east of the crater Julius Caesar, when illuminated near the terminator.

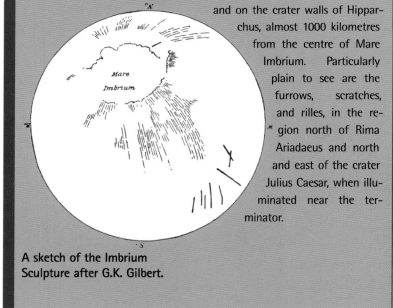

A sketch of the Imbrium Sculpture after G.K. Gilbert.

Lacus, Palus and Sinus

Less extensive lava flows are known by the Latin terms Sinus (Bay), Lacus (Lake) and Palus (Swamp). In many of them their origin is uncertain: whether we are dealing with lava-covered surfaces linked to the large maria, or whether they arose through individual impact events. Notable examples are:

- Sinus Iridum (Bay of Rainbows), where it is definite that here there was an impact before the formation of the Imbrium Basin. In the subsequent Imbrium impact the whole of the southeastern half of the wall of Iridum was flooded by Imbrium lava.
- With Lacus Timoris, its origin is uncertain. The lava surface is elongated and lies in the southwestern highlands. In the centre it is constricted by cape-like mountains and it could also be interpreted as two, lava-flooded large crater remnants.
- Palus Epidemiarum is an irregular lava flow, broken by numerous cracks and fracture zones, which might reasonably be linked to the lava of Mare Nubium. It lies in the Moon's southwestern quadrant.

Dorsum and Dorsa

The Latin word Dorsum is applied to what is known as a mare wrinkle ridge; Dorsa is the plural and is used to describe a complex system of such structures. The mare wrinkle-ridge systems lie on the major basalt lava flows that once flooded the large lunar basins.

Wrinkle ridges are the most striking features in the widespread and otherwise completely flat mare surfaces. Because of their very low height of about 100 m to 250 m as a maximum, they may only be seen under very low solar illumination, but are then very striking. When the Sun is high over the region in question, they become completely invisible. This slight difference in height might be likened on Earth – albeit at a very small scale – to the ripples that are produced on fine beach sand by the action of the waves.

Many of the mare wrinkle ridge systems follow the arcuate shape of the inner edges of the maria, as, for example, on the eastern side of Mare Humorum. In Mare Crisium they run near the border – almost completely encircling it – inside the whole mare surface. The average height of the Mare Crisium wrinkle ridges amounts to only 100 to 150 m. In Mare Serenitatis (and in other maria as well) wrinkle-ridge segments also run in approximately linear fashion. For example, portions of Dorsa Lista and Dorsa Smirnov are mostly oriented north-south.

If the individual segments of the wrinkle ridges in Mare Imbrium are viewed as a whole and in relation to the mountain ridges and isolated mountains, then it is possible to see that, on a large scale, they form a ring-shaped structure with a diameter of about 650 km. This circle lies concentric with the outer basin rim of Mare Imbrium, recognizable in the Apennine mountains. The ratio between the Mare Imbrium wrinkle-ridge system with a diameter of c. 650 km and the outer basin rim amounts to exactly 1:2, and similar ratios are found in almost all maria, between the wrinkle ridges and the outer basin rims.

If these ridges are observed in detail, it is found that their structure is not symmetrical. On one side, the slope from the broad bulge of the crest is very slight, whereas the other side is distinctly steeper, casting a shadow on the surrounding lava surface. The different slopes frequently, and repeatedly, switch from one side of the crest to the other – for no apparent reason – so that the ridge appears somewhat like a twisted rope.

Mare wrinkle ridges probably arose when the lava flows that flooded the basins petered out. The central area of the lava pile began to sink under its own enormous weight, assisted by the fact that the pressure of the rising lava from the mantle declined. The basalt in the outer regions of the mare slid inwards, was compressed and folded into the wrinkle ridges. At the same time the linear faults and graben were formed in the outer borders between the mare and the highlands, where the basalt layers were thinnest. Probably both effects overlapped and the folds in the basalt through the sinking of the mare surface occurred along the inner basin walls of multi-ring basins.

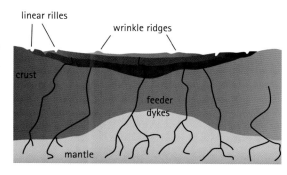

An idealized section through an impact basin that was flooded by three successive lava flows. After Hiesinger, 1999.

Terrae

The early lunar observers such as Riccioli and others saw the highlands – in contrast to the maria – as continents and thus designated them 'terrae'. In them we indeed see the Moon's original silicate crust that crystallized from the original magma ocean, which covered the whole surface of the Moon. There creation has been dated to a time before about 4.1 to 4.3 billion years.

Mountain ranges and single mountains

The mountain ranges on the Moon cannot be compared in their origin with mountain ranges on Earth. The major mountain ranges on Earth have arisen though plate tectonics and have subsequently been altered through wind and water erosion, to the state that we see today. On the Moon there has probably never been any plate tectonics. The Moon's mountain ranges are the remnants of crater rims that were thrown up by the major impacts, when the basins were formed. They are highest portions of those crater rims, which were not submerged by the mare lava. The enormous seismic pressure waves, released by the impact, heaped up the partially and completely molten lunar crust in the form of waves that were concentric to the centre of the impact into extremely high crater rims.

Wrinkle ridges in Mare Crisium (left) and Tranquillitatis (right).

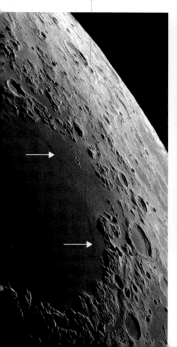

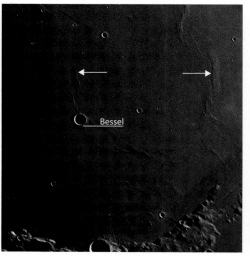

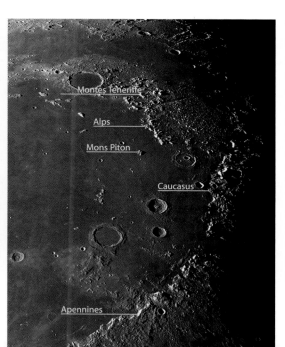

The three huge mountain chains that encircle the eastern portion of Mare Imbrium: the Apennines (Montes Apenninus), Caucasus (Montes Caucasus) and the Alps (Montes Alpes), are the remnants of a rim around the Imbrium Basin. The same applies to the isolated mountains (e.g., Mons Piton) and the short mountain ranges (Montes Teneriffe, Montes Recti).

The inner walls of these basin rims (on the mare side) are, with their slopes of c. 30°, similar in steepness to normal craters. The outer surroundings (the hinterland) is widely covered by large quantities of ejected material and blocks, and the angle of the flanks, at 5 to 15°, is relatively flat.

Over the course of hundreds of millions of years, erosion has rounded and flattened the once sheer upper slopes of the rims. The most impressive example of little-eroded basin rims is Mare Orientale, with three, almost complete, encircling crater walls.

Rupes

The Latin word 'rupes' may be taken to mean 'furrow', 'steep escarpment', 'slope', and even 'cliff'. Most of the surface features designated rupes are portions of the rims of impact basins. So, for example, Rupes Altai is a segment of the rim of the Nectaris Basin. In its appearance it resembles an escarpment or cliff. The same applies to Rupes Kelvin and Rupes Liebig, that are certainly the remnants of a rim of the Humorum Basin. Rupes Mercator is similarly a remnant of the wall of the Nubium Basin. They all have relatively steep slopes in the direction of the basin.

Two further features on the nearside of the Moon that are also given the designation 'rupes', are Rupes Recta (in the eastern portion of Mare Nubium) and Rupes Cauchy (lying in the northeastern part of Mare Serenitatis). Both are so large, that they can be well observed under grazing solar illumination even in smaller telescopes. They are both large-scale ground subsidences and thus arose under different geological circumstances than the rims of the impact basins. Both subsidences are also associated with linear rilles that run parallel to them (Rima Birt and Rima Cauchy).

Rupes Recta and Rupes Cauchy both have a length of c. 120 km. The difference in height between the downfaulted side and upper edge of the escarpment amounts to just a couple of hundred metres (about 250 to 300 m in the case of Rupes Recta) over the slope's width of 2 to 3 km. This gives an inclination of less than 10°, and so both are fairly gentle slopes. Because both structures lie in the border regions of large maria, it may be assumed that their formation could be related to the mare lava. It is conceivable that such subsidences in unstable areas of the lava blanket may have collapsed as a result of seismic shock waves produced by the Imbrium impact, and have been modified by later lava flows. Rupes Recta and Rupes Cauchy are fairly closely aligned towards the centre of the Imbrium Basin, and also when it comes to age, the Imbrium Basin was created after the Serenitatis and Nubium Basins.

Rupes Cauchy is, in this respect, a somewhat peculiar feature, because the end of the slope turns into a small rille system. (A similar structure may be observed in the southern area of Lacus Mortis with Rimae Bürg.) Near both regions there are signs of lunar volcanism. Near Rupes Cauchy this is seen in the large number of lunar domes. With Rupes Recta, the rille, Rima Birt, which runs parallel to it, shows clear signs of active volcanism. Rima Birt is flanked at both ends by small cratered peaks, around which dark material has been deposited. So Rima Birt is one of the few exceptions, where a linear rille lies within a mare surface and which may have arisen through active volcanism.

Rupes Toscanelli, the remaining formation on the nearside of the Moon, which according to the old lunar nomenclature was designated 'rupes', is probably the opposite of a subsidence – an elevation of the ground, possibly also caused by the Imbrium impact. It is on the eastern side of the Aristarchus plateau, and here

Domes are bulges of volcanic origin, as here with the domes near the crater Gruithuisen.

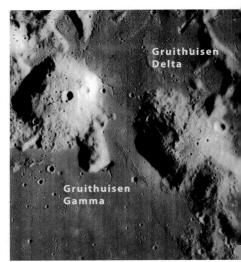

the inclination or gradient is significantly greater than with Rupes Recta and Rupes Cauchy.

Domes

In accordance with lunar nomenclature, lunar domes were generally described as mountains. From a geological point of view, however, Mons Pico and Mons Gruithuisen or Mons Rümker are completely different structures.

Lunar domes are undoubtedly of volcanic origin, and of two different types. For example, the two Gruithuisen domes are so-called intrusive domes. They consist of a very rare silica-rich volcanic rock, that has crystallized from an olivine- and pyroxene-rich basalt lava, as determined by multi-spectral investigations (which are also able to show that central peaks consist of minerals that were formed at depths of many kilometres).

These rocks were slowly elevated and bowed upwards by the intrusion of molten lava between existing layers. With intrusive volcanism there is no extrusion of lava onto the surface. Purely intrusive volcanism is relatively rare, because in most cases it turns into so-called extrusive volcanism, in which fluid lava spreads over the surface.

Some examples of purely intrusive volcanism on Earth are the Novarupta Dome in Alaska (with a diameter of around 2 km and a height of about 800 m), Paoha Island (in Mono Lake, USA), and the famous Devils Tower in Wyoming, USA. The latter is relatively old and the dome has been eroded, such that just the core remains. Paoha Island is different: it is only about 300 years old, and lake-floor sediments have been raised by the intrusion.

The other group of lunar domes are most closely comparable with shield volcanoes on Earth. They do occur individually, but are generally found in larger groups. Individual domes are mainly found on the floor of the larger craters or complex craters. Groups of lunar domes are often found in mare areas and frequently association with areas of the so-called Dark Mantle Deposit (DMD).

Domes have, on average, diameters of 5 to 20 km, whereas their heights generally amount to just a few hundred metres. Correspondingly, the slope gradients are only a few degrees. Many lunar domes have a central, summit crater, a caldera, of about 1 km in diameter. A very few have two or even more such summit craters. Mostly, they are round in shape. The largest, which are visible even in small telescopes, are the domes Arago Alpha and Arago Beta in Mare Tranquillitatis and the dome Kies Pi, directly west of the crater Kies in Mare Nubium.

Larger groups lie directly north of the crater Hortensium and south of the crater T. Mayer in Mare Insularum, west of Copernicus. A large group lies northwest of Rupes/Rima Cauchy. The largest group of associated domes –

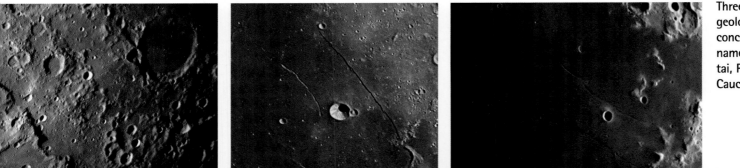

Three completely different geological processes are concealed in three similarly named features: Rupes Altai, Rupes Recta and Rupes Cauchy.

An example of a superdome is the unofficially named formation 'The Helmet' near Gassendi. The northern crater of Ol Doinyo Lengai in Tanzania shows a terrestrial example of how one may envisage a superdome appearing in detail.

the 'Marius Hills' – lie west of the crater Marius in Oceanus Procellarum. They are also associated with very small, sinuous rilles (not visible in amateur telescopes), which are also of volcanic origin. Modern, precise height measurements by lunar probes lead us to suspect that the Marius-Hills region, and the domes in the Cauchy and Hortensius regions lie on gigantic shield volcanoes.

Megadomes and plateaus

'Megadomes' are defined as flat, plateau-like elevations on the lunar crust, which are mostly of approximately circular outline. Their diameters lie between 30 and 50 km and at a few hundred metres, their heights correspond to those of normal domes. Their surfaces seem to be shallow convex bulges and they carry small, shallow domes or dome-like structures, and sometimes small craters, rilles and peaks. The plateau-like rises may be of volcanic origin, created over a long period of time through the pressure of rising fluid mantle lava and which then slowly declined from the small summit craters.

Crustal uplift caused by a large, nearby impact is also conceivable, where the surface was elevated to such an extent that it was not flooded by the subsequent lava flows. Such a crustal uplift occurs in a very short space of time, shortly after the impact event. The most striking megadome plateau is Mons Rümker in Sinus Roris on the northwestern limb of the Moon. Two others have no official name, but have been nicknamed 'The Helmet', east of the large crater Gassendi, and 'Valentine Dome', east of the southern end of Montes Caucasus. In all three cases, their morphology appears to suggest that they formed through uplift of the crust by lava. The plateau south of the crater Gardner on the borders of Mare Serenitatis and Mare Tranquillitatis is also a candidate for being considered an uplifted megadome plateau. Yet another megadome lies directly north of the crater Milichius.

A different form of uplift applies to the Aristarchus Plateau, where the region was a low-lying surface in a former highland area, which was raised by the gigantic impact that created the Mare Imbrium. The many sinuous rilles and the lunar domes in the vicinity also show, however, that this region – at least in part – has been strongly affected by volcanism. The difference in height between the surrounding landscape and the top of the plateau is a magnitude greater than with normal superdomes.

The two variant methods of formation are theories, not conclusively resolved, and must undoubtedly be evaluated for each megadome individually. Pyroclastic, strongly eruptive, lunar volcanism in the form of volcanic ash and dark lava flows associated with rilles and craters may be observed (for example) on the floors of the craters Alphonsus and Atlas.

Many instances of lunar volcanism resemble features known on Earth. In Iceland, in particular, many of the small-scale structures are found and it was for this reason that NASA trained the astronauts who flew to the Moon or who might have done so, in the area around the Hverfjall cinder cone near the Myvatn Lake.

Rima, Rimae

The classification 'rima' is generally known as 'rille', although it may apply to a rift, crack or graben. The plural 'rimae' is often used to designate a whole complex of rilles. Lunar rilles are sub-divided according to the course, which distinguishes them as sinuous or linear rilles or rille systems. Sinuous rilles, which meander through their surroundings are found almost exclusively in mare areas. They show an astounding similarity to dry, terrestrial river courses. They may exhibit abrupt breaks and often split and branch out. In contrast to terrestrial rivers (which broaden from source to mouth), sinuous rilles narrow before they end downstream. They often begin near a small crater.

Following intensive investigation of the Hadley rille by the Apollo-15 crew, lunar geologists have interpreted them as tube-like channels in which lava once flowed. After volcanic activity died away, and the flow of lava ceased, the roofs of the tubes collapsed, leaving a rille, or open channel. Such lava tubes are well-known on Earth in areas of volcanic activity. Sinuous rilles are also found on the planets Venus and Mars.

Impressive examples of sinuous rilles are Rima Marius, Rima Herigonius, Rima Hadley and Vallis Schröteri on the Aristarchus Plateau. Vallis Schröteri is the largest and, at its source, the widest, and is thus easy to observe with small telescopes. In the lunar nomenclature Vallis Schröteri has been given an in-

Rima Hadley is one of the most famous examples of a lunar rille. Corresponding features on Earth are lava tubes, as here on the Galapagos Islands.

correct classification, because it is undoubtedly a sinuous rille and not a valley-like structure.

Straight or slightly arcuate rilles are primarily found on the borders of highland regions (running fairly straight), and in the border areas between highland and maria (where they are mostly arcuate). They are described as being linear rilles or rille systems. The arcuate rilles frequently follow the course of the walls, surrounding the lava-flooded areas of a mare. Most of them are faults in fracture zones that arose through the slow contraction of the central lava flows in maria. The lava layers in the central areas are significantly thicker than around the edges, so it is in these unstable border areas that fractures and faults occur. When two such faults slowly move apart, either through tension stresses or shear movements, the ground between them collapses producing a trough or valley. One example – visible in small telescopes – is the Rimae Hippalus system in the eastern part of Mare Humorum. Here tension in the cooling and thereby shrinking mass of lava has created three, parallel, broad, valley-like troughs. In rare cases, however, linear rilles are clearly associated with volcanism, and are probably of volcanic origin. Examples of this are Rima Hyginus (on the boundary between Mare Vaporum and Sinus Medii), the narrow rilles in the crater Alphonsus and Rima Birt in Mare Nubium. Most rilles of both types are, on average, only a few kilometres wide and their depths amount to just a few hundred metres.

Vallis

The Latin word 'vallis' means 'valley' and here again early lunar observers were in error, in that they likened valleys on the Moon with terrestrial valleys. Whereas valleys in Earth's mountainous regions were formed by flowing water and were carved into the structure of mountains, almost all lunar valleys were created by impact events. The only two exceptions are Schröter's Valley (Vallis Schröteri, a sinuous rille) and the Alpine Valley (Vallis Alpes, a graben trough).

The Alpine Valley consists of two, parallel fracture zones, between which the lunar surface has collapsed, actually forming a typical lunar rille, albeit one of gigantic size. The Alpine Valley probably arose at the same time as the Imbrium impact, and the floor of the valley was slowly filled and smoothed by lava flows. Similar areas may, in principle, be found on Earth, although here arising from continental drift through plate tectonics. Examples are the Thingvellir (Þingvellir) valley in Iceland and the Rift Valley in East Africa, on the border between Kenya and Tanzania.

All the other formations that may be observed on the nearside of the Moon and which are classed as 'vallis', consist of more-or-less strongly eroded crater chains (catena) made up of partially superimposed craters. It is not possible to distinguish whether individual valleys arose through impact events – similar to the formation of chains of craters – or through secondary impact events; this cannot

be determined without further investigation. What is striking, however, in many of these valleys, is their radial direction relative to the major impact basins, as with Vallis Capella to Mare Serenitatis/Mare Tranquillitatis and Vallis Rheita to Mare Nectaris. An example of the major crater chains on the Moon's farside is Vallis Bohr, that lies radially to the centre of Mare Orientale.

One exception is the longest lunar valley, Vallis Schrödinger. It practically cuts the crater Sikorsky in two, and is oriented radially to Mare Australe. Here very few individual craters are visible, and so the formation of Vallis Schrödinger may resemble that of the Alpine Valley, it having been flooded with lava from Mare Australe. The individual craters that combine to form Vallis Snellius (south of Petavius) are already heavily eroded. Vallis Snellius is orientated radially to the centre of Mare Imbrium and is probably part of the Imbrium Sculpture.

Craters

The concept of lunar craters is so widely accepted that, following traditional selenographical literature, craters and their description on maps are roughly differentiated by their size and appearance, even though this classification is not officially accepted nowadays in modern impact research. On the Moon's nearside there are about 300 000 craters with a diameter greater than 1 km, and 234 of those are greater than 100 km in diameter. On the Moon's farside there are significantly more. From this it may be concluded that, on average, the diameter of a crater is about 10 to 20 times larger than the diameter of the impacting body. A crater of the size of Copernicus, with a diameter of just under 100 km was therefore produced by an impacting body of 5 to 10 km in diameter.

In classical lunar literature, the term 'walled plains' was used for relatively flat craters with diameters of between 60 and 300 km. Their crater rims are often wide and full of structure. The formerly sharp ridges have generally been degraded by later, smaller impacts, and partially destroyed, as well as being rounded by the ejecta from subsequent impacts. In the largest of the 'walled plains' the floor of the crater is convex, following the curvature of the Moon. Frequently, smaller craters, formed later, lie on the floor of the crater. Hills and the remnants of central peaks are visible, and the floor of the crater is often lava-flooded, flat and smooth. Because of their size these craters are frequently significantly older than smaller craters. Particularly interesting examples are Clavius and Ptolemaeus, among others. Structures that are larger than 250 to 300 km are classified as lunar basins. An example is Grimaldi on the Moon's western limb.

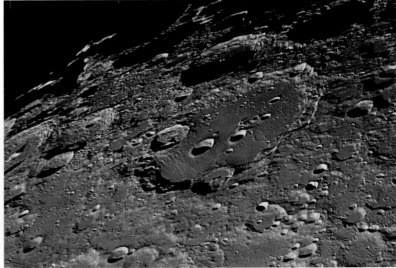

Clavius is a classic walled plain.

Craters with diameters between 20 and 100 km are described as 'ring mountains' or 'mountain rings'. The crater ramparts are often sharp-edged and little eroded. The shape of the crater is approximately round and symmetrical, and the crater upper rim often consists of six to eight linear segments that form a circle. As a rule, the floor of the crater is lower than the surrounding terrain. The inner walls of the crater are often terraced, and images from lunar probes show clearly that

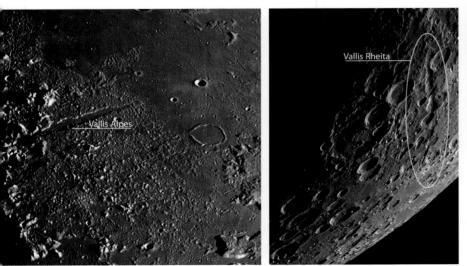

Both formations are described as 'valleys', but Vallis Alpes is a graben trough, and Vallis Rheita a crater-chain.

these terraces originated as landslides. The gradient of the inner crater walls (20° to 30°) is significantly steeper than that of the outer rampart, where the gradient amounts to just 5 to 15°. Ring mountains almost always show a prominent central peak or complex central peaks. The base of the crater is frequently rough. Ring mountains are significantly younger than the walled plains, and are recognizable by the lesser damage to the crater walls through subsequent impacts. Good examples are Copernicus, Tycho and Theophilus.

True craters are described in the classical lunar literature as circular depressions, surrounded by crater walls, with diameters between 5 and 60 km. The crater walls are sharp-edged, the inner crater walls are rarely terraced and, at diameters less than about 30 km, they exhibit no central peak or peaks. At the larger diameters the floor of the crater is flat and smooth. Smaller craters may even be funnel-shaped and, under certain lighting conditions may even appear as if a cone had been forced into the lunar surface. An example of the larger craters is Kepler. Examples of conical craters are Kies A, B and E, and Birt, among others.

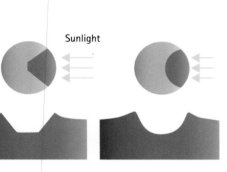

Sunlight

Simply from the form of the shadow of the crater wall, it is possible to determine whether the floor of a crater is flat or bowl-shaped. A bowl-shaped floor exhibits a semicircular shadow, whereas a flat floor with straight inner crater walls produces a shadow that is triangular. After Wood, 2007.

Depressions smaller than 5 km in diameter are designated craterlets or pits. They are round and symmetrical in shape, and are probably to be seen as primary impacts (for example, Linné). Elongated or irregularly shaped craterlets are mainly secondary craters. They occur following an impact, when the impact energy has broken up the crustal material, flung it upwards and explosively hurled it outside the crater. The blocks of rock are generally partially or completely molten and the angle at which they are projected is very flat, which leads to the formation of irregular or elongated craterlets. Impressive examples are the surroundings of Copernicus, and particularly the area around the crater Stadius, or the northeastern area around Theophilus, which are saturated with secondary craters.

If relatively young craters such as Copernicus or Tycho are seen when the illumination is near the terminator, then because of the sharp, densely black shadows cast by the crater walls it produces the impression that they are 'bottomless', sheer pits in the lunar crust, with steep, needle-sharp, towering central peaks. However, if one takes the ratio of the crater's diameter to that of its depth (or the height of the walls) then it becomes obvious that the crater is in the form of a flat-floored depression rather than a deep hole. Crater depths or wall heights and the heights of the central peaks have been determined from the length of their shadows and the elevation of the Sun above the lunar surface and are now well-known since the Lunar Orbiter missions.

The larger the diameter of the crater, the greater the ratio it bears to the depth. The average value for walled plains is about 1:64 (depth versus diameter, such as in Ptolemaeus). Mountain rings, such as Copernicus show a ratio of about 1:25. The crater Kepler has a diameter of 32 km and a depth of about 2.7 km, which corresponds to about 1:12. With even smaller craters, such as Hortensium, for example, the

Relief in the crater Tycho, as imaged by the Clementine Moon probe.

diameter is 14.6 km, and the depth 2.8 km. That corresponds to a ratio of 1:5 and gives the appearance of a deep hole in the Moon's surface. The same applies to the central peaks in the large mountain rings. To visual observers they appear steep and abrupt but in reality they are gently rounded, relatively flat, conical structures, as, for example, is shown by the image of the central peak in Tycho that was returned by the Clementine probe.

The standard sequence of crater morphology

Simple craters
There are thousands of examples of the simple type of crater on the nearside of the Moon. High-resolution images from lunar probes show that the inner walls of these craters are relatively steep, and that the crater rim rises only a few dozen metres about the surrounding terrain. Most of them are conical in shape, down to the full depth of the crater floor. A few, however, also exhibit a flat floor, depending on whether loose masses of rock have slid down the inner crater walls after the impact.

Complex craters
Craters with diameters greater than 20 km exhibit a completely different morphology to simple craters. Examples of complex craters are the 26-km-diameter Triesnecker, with a somewhat irregular shape to the crater wall, or the 25-km crater Euler. The rock in the uppermost layers of the crater rim has a high albedo. On the floor of the crater there is a collection of rocky rubble and boulders, that has obviously fallen from the inner wall. A 'bite' seems to have been taken out of the western wall of Triesnecker, which is the result of a massive landslide, where a portion of the inner crater wall has slipped. The rock has spread out on the floor of the crater. In the middle of the crater's floor there is a complex central peak. A flat, smooth, mare-like structure, consisting of melted ejecta, covers the crater floor.

Many craters with diameters of between 20 and 40 km show such massive landslides, which often appear polygonal. It is notable that the landslides are not distributed fairly equally around the inner crater wall, but always lie to one side. About 80 such craters may be observed on the Moon's nearside. While simple craters have, as a rule, a depth that is about 1/5th of their diameter, complex craters – because of the massive landslides – have a depth/diameter ratio of 1:10 to 1:40; they are thus significantly shallower, and in cross-section resemble a frying-pan, rather than a bowl-shaped cavity. In addition, the floors of complex craters are filled with melted ejecta, which cooled rapidly and thus gave the floor its flat, smooth surface.

Most simple, small-sized, lunar craters have a morphology that is conical in shape.

The morphology of the 17-km-diameter crater Bessel exhibits a transitional phase. The structure of the crater's floor now resembles that of a complex crater, but the central peak is missing.

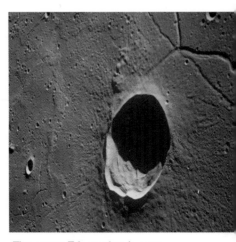

The crater Triesnecker has an irregular outline, and has been affected by landslides.

Complex craters arise because of an interaction between the strength of the rocks of the inner crater wall and the Moon's gravity. In simple craters, the composition of the rock – during the second phase of the impact – is firmer and the lunar gravity is not strong enough to cause the rock of the inner wall to collapse onto the floor of the crater. In complex craters the central peaks are formed at the same time as the collapse of the inner crater walls. Initially, the lunar crust at the centre of the impact is forced downwards into the Moon's interior by the immense pressure. When the pressure is removed, the compressed rock rebounds and is raised above the level of the crater floor in a heap. (A terrestrial analogy is perhaps found in images of water droplets taken with high-speed cameras.) The rebounding rock is probably partially liquefied by pressure and heat so that it is viscous.

Studies of terrestrial impact craters with central peaks have shown that there is a close relationship between the crater diameter and the height of the central peak – and with it a relationship between the diameter of the impacting body and its impact velocity.

Large complex craters

Copernicus, with its diameter of just 96 km, is about four times the size of Triesnecker. Its inner ramparts are in the form of terraces, the crater floor is flat, smooth, and largely covered in flat hills. The central mountain range is complex, with numerous peaks. The structure of craters changes around diameters of 35 km from the Triesnecker to the Copernicus/Tycho type.

Crater size and height of central peak	
Crater diameter	Height of central peak
20 km	0.5–1 km
40 km	1 km
60 km	1.5–2 km
80 km	2 km
100–140 km	2 km and higher

Copernicus is the perfect example of a large complex crater.

The crater Schrödinger is a good example of the peak-ring crater type, but is, however, not visible from Earth.

In craters of this size, the landslips no longer consist of more or less loose rubble. Instead, the inner walls of the crater sag under their own weight into multiple annular terraces making up a series of steps. The pressure and temperature possibly caused the rocks at the impact point to behave as a viscous fluid. As such, the material did not slump vertically, but slid over a curved surface towards the centre of the crater. This increased the crater's internal diameter.

Above a crater diameter – crater-wall diameter – of 150 km, a further change in structure become evident. Individual central peaks (or ranges of peaks) become rarer (and also smaller) and are replaced by a circular zone of irregularly located peaks, concentric with the crater's centre. In the specialized literature they have been called 'peak-ring craters', and this transitional stage is also known from craters on Earth, Mercury and Mars. Such craters are not observed on the icy satellites of Jupiter and Saturn. The diameter of the ring zone is, on average, about half that of the crater's diameter. The origin of ring of mountains is thought to be related to the size of the molten zone beneath the impact zone. Unfortunately, there is no striking examples of such peak ring craters that may be observed on the nearside.

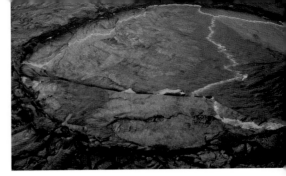

An indication of how an FFC-crater may have appeared billions of years ago is given by the Danakil Depression in the Ethiopian rift valley.

Basins and multi-ring basins

Basins are the largest impact craters on the Moon and their formation and the conversion of the enormous energies that were released defy imagination. Large complex craters with diameters of 250 to 300 km are defined as basins. The largest is the South Pole-Aitken Basin on the farside of the Moon, with a diameter of about 2300 km and a difference in height of up to 16 km. Isolated central peaks are no longer present, and the annular zone of irregular mountain peaks in the peak-ring craters has become a continuous external crater wall. The lunar crust was so greatly destroyed and so deeply fractured that the centres have been flooded with mantle lava. Grimaldi, Schickard and Mare Humboltianum are examples of smaller basins.

The transition in morphology between basins and multi-ring basins lies in the fact that the latter display multiple concentric, mountainous rings. The complex basin structures are now hardly visible on the nearside of the Moon, because the central regions have been largely flooded by lava and the basin walls have been destroyed by later impacts.

The most impressive example of a young (and probably the youngest) largely unflooded basin is Mare Orientale, only portions of which are, unfortunately, visible from Earth, and then only under favourable libration conditions. The best images come from Lunar Orbiter 4, which made numerous passes directly over the basin in 1967. The Mare Orientale was discovered as early as 1935 through the study of lunar structures near the limb. The first television pictures of the Moon's farside from Luna 3 were so blurred that only the dark central lava surface of the basin was visible.

One theory for the formation of multiple concentric crater rings (similar to the ripples that form on the surface of water after the impact of a stone), is based on the idea that the enormous pressure of the impact on the lunar surface renders it largely viscous and the expanding shock waves heap up the surface in waves. The physical and geological processes that take place are still poorly understood and are the subject of current research.

Different forms of craters

Almost 99 per cent of all lunar craters essentially correspond, in their structure and morphology to the standard crater sequence. Most of the craters with different morphologies may be ascribed to later modifications to the craters. Only the polygonal and elliptical craters have to be accounted for by special circumstances accompanying the impacts.

FFC-craters

Complex or large complex craters in the standard sequence either have a lava-flooded, relatively flat and smooth floor, or the floor is chaotic and hummocky and has been affected by landslides. The depth of the crater is essentially dependent on the diameter: the larger the crater, the shallower it is.

The abbreviation FFC stands for 'floor fractured crater'. This is applied to craters where the crater floor exhibits a complicated system of fractures and rilles. In some cases (e.g., Pitatus) these fracture zones follow the course of the inner crater wall, but in other cases may run all over the crater floor (e.g., Gassendi). Both forms may occur in a single crater (e.g., Atlas).

Apart from the rille systems, all FFC craters exhibit two further notable common features. First, they all lie on the borders of maria and, second – in contrast to normal complex craters – the depth of the crater floor below the wall is extremely low. The craters Sabine and Ritter on the southwestern border of Mare Tranquillitatis are, for example, only about 700 to 750 m deep, whereas, based on their diameter, they ought to be about 2 to 2.5 km deep.

Nowadays it is firmly believed that the FFC craters were produced by a normal impact and that formerly the crater floor lay much deeper. According to the current theory among lunar geologists, at some time – long after their formation – lava from the mantle spread through fracture zones beneath the basins, was forced into cracks beneath the crater and slowly raised the whole crater floor. Because the pressure of the rising lava was not precisely the same everywhere, cracks formed in the crater floor.

Gassendi is an example of an FFC crater that is filled with lava and crossed by rilles.

The annular fractures along the inner crater wall probably formed after the pressure of the formerly rising lava declined, and the whole crater floor sank back slightly under its own weight. Because the lava layer was thinner at the edge, the fractures occurred there.

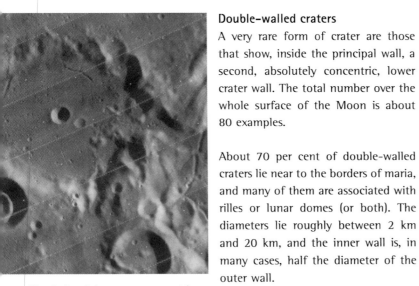

Hessiodus A is a rare, concentric, double crater.

Double-walled craters

A very rare form of crater are those that show, inside the principal wall, a second, absolutely concentric, lower crater wall. The total number over the whole surface of the Moon is about 80 examples.

About 70 per cent of double-walled craters lie near to the borders of maria, and many of them are associated with rilles or lunar domes (or both). The diameters lie roughly between 2 km and 20 km, and the inner wall is, in many cases, half the diameter of the outer wall.

Lunar geologists believe that the double walls did not arise through two impact events, occurring one after the other, but ascribe the formation of the outer wall to the impact, and that of the inner wall to the intrusion of highly viscous magma in an annular zone.

For amateurs, the double-walled crater that is simplest to observe is Hessiodus A on the southern edge of Mare Nubium, which has a diameter of 14.9 km and a depth of 1.7 km. 70 per cent of all double-walled craters resemble Hessiodus A.

Under low illumination, Hessiodus A appears like a normal crater. When the elevation of the Sun increases and illuminates the interior of the crater, the conspicuous double structure becomes immediately visible. In the centre of the inner wall there are two or three small central peaks. The crater shows a slight ray system.

Elongated and elliptical craters

True elongated or elliptical craters are even rarer on the Moon than double-walled craters. Most of them are actually nearly round, and only appear elliptical because of limb foreshortening.

Experiments by NASA have shown that elliptical craters can occur only if the angle of the impactor's path to the surface is less than 3° and if the impact velocity is very low. So most elliptical craters have arisen from secondary impacts.

On the southwestern limb of the Moon two large elliptical craters may be seen: Hainzel and Schiller. Hainzel very clearly consists of three overlapping impact craters, where parts of the separate crater walls are still visible.

Schiller, with a width of c. 70 km and a length of 180 km, is highly elliptical. Its interior is completely free from any dividing structures and the crater floor appears very smooth. Crater counts suggest an age of about 3.7 billion years. Experiments by D.E. Gault have established that a grazing impact produces several elliptical, overlapping craters. Schiller may perhaps be the visible remains of an earlier satellite of the Moon. An earlier hypothesis, that Schiller is a gigantic, collapsed volcanic caldera similar to the 100-km long and 35-km wide Toba Caldera in Indonesia, nowadays appears unlikely, based on the terraced inner walls and the complete lack of dark ash deposits surrounding Schiller.

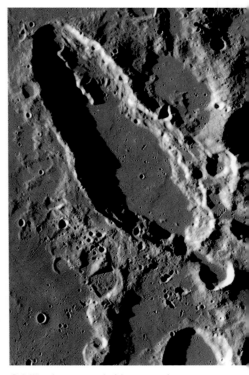

Schiller is a remarkable example of an elliptical crater.

Polygonal craters

While the majority of all crater walls are nearly round – or appear elliptical to observers because of perspective effects – a few craters exhibit crater walls that are more or less polygonal. A striking example is Proclus on the western edge of Mare Crisium. To date, there is no conclusive explanation for the formation of such craters. It is suspected that there is possibly a relationship to the impact angle and the velocity of the impactor. Frequently, young polygonal craters exhibit a very asymmetric ray system, which results if an impacting body hits the surface of the Moon at a relatively low velocity and at an angle of less than 5°.

Two further examples of polygonal craters are Calippus and Theatetus, which lie directly southwest of Callipus. Calippus lies on the crest of Montes Caucasus and Theatetus on the western edge of the Caucasus on the Imbrium lava. Neither crater has a ray system, but both have been dated as secondary craters from the Imbrium impact.

If they are truly secondary craters, they will also have arisen from a slow projectile velocity and a low angle of impact. If we were to follow the rules of stratigraphy, however, Theatetus could not be a secondary crater – at least at first sight – because it lies in the Imbrium lava and must have arisen long after

Kepler and Encke both have a distinct pentagonal shape.

flooding of the basin took place. However, if the crater is observed under very low illumination, then it may be seen that it lies on a very broad, linear elevation, the surface of which is significantly rougher when compared with the smooth Imbrium lava. Given its higher location on a ridge (part of the outer basin ramparts), Theatetus may – on second thoughts – indeed be a secondary crater from the Imbrium impact, whose surroundings have been covered by just a very thin layer of Imbrium lava.

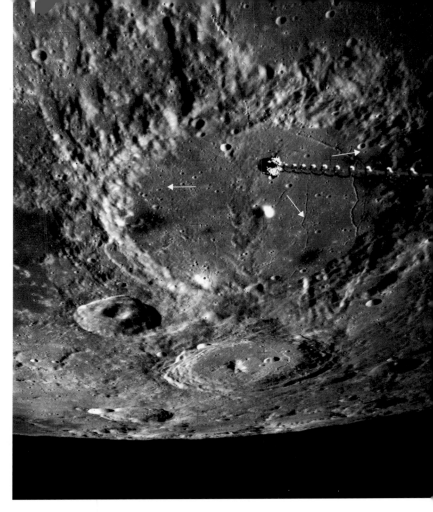

The dark-halo craters in Alphonsus, as seen from Apollo 15. The image also clearly shows the difference between the central peaks of Alpetragius and Arzachel (bottom).

Craters with abnormal central peaks

A few craters have central peaks that differ from the normal, slender pointed shape. They appear broad and rounded, with wide bases. The best example for observation if the crater Alpetragius (40 km diameter and 3.9 km deep), directly southwest of Alphonsus. Alpetragius undoubtedly has the most extraordinary central peak of all lunar craters. Its base fills the whole of the crater floor. In the wider terrain around Alpetragius such craters occur in great numbers. Among others there are Parrot C, Faye, Donati and Airy.

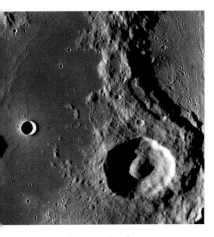

Alpetragius has an unusual, round central peak.

The departure from the norm that exists in Alpetragius becomes all the more striking if it is compared with the crater Herschel, directly north of Ptolemaeus. The latter has nearly the same diameter and depth and corresponds, in the USGS classification, to a complex crater with a small central peak, flat crater floor, and terraced inner crater walls.

Central peaks normally form through the rebound of the lunar surface, depressed by the pressure of the impact, and they cover a maximum of 30 per cent of the crater floor. The central peak in Alpetragius was possibly altered to give its current form by later eruptive lava flows within the crater. This theory is tenable, because the floor of Alphonsus was also altered a long time after the impact by subsequent eruptive volcanism. Craters of the Alpetragius type are extremely common on Mars.

Dark-halo craters

Normally, young impact craters show bright rays systems that strongly reflect sunlight, or round haloes derived from the anorthosite of the upper lunar crust. Occasionally, however, we come across what are generally small craters that are surrounded by a halo of dark material. They are also particularly clearly visible under high solar illumination, and even more prominent if they lie on top of the ejected material from bright ray systems. They are known as dark-halo craters (DHC).

There are two different classes of DHC craters; one consists of normal impact craters, and the other are undoubtedly of volcanic origin.

If the area around Copernicus is observed under high illumination, many small craters stand out that are surrounded by haloes of very dark material. The most prominent is Copernicus C, which has a diameter of about 5 km, lying southeast of Copernicus. The halo itself has a diameter of about 25 km. Multispectral investigations unequivocally confirm that the halo consists of pulverized mare lava. The impact that created the crater, penetrated deep enough below the ejecta from the Copernicus impact, to lift the underlying basalt lava up to the surface. Copernicus, a young crater, arose long after the lava flows.

The best examples of dark-halo craters of volcanic origin lie on the floor of the large crater Alphonsus. In their general morphology they differ little from the dark-halo impact craters. They are, however, always associated with cracks and rilles. The dark haloes around these craters consist of volca-

nic ash from strongly gas-enriched lava, which erupted explosively (similar to the eruptions that have spread ash over the regions with dark mantle deposits). The craters in Alphonsus are only 2 to 3 km in diameter. The haloes are larger, however, and thus easier to observe. Further examples of these small volcanic craters lie on the floor of Atlas, here again associated with cracks and rilles.

Catena

In lunar nomenclature the Latin word 'catena' is used for a chain of craters. On the nearside of the Moon there are some extremely regularly arranged chains of craters that may be observed. The best-known are Catena Davy and Catena Abulfeda. In the last century they were indeed put forward as the principal argument for the volcanic origin of lunar craters. Such chains of volcanoes and their catastrophic eruptions were known on Earth, where, for instance, an eruption of the Laki chain in Iceland in 1783 almost led to the complete evacuation of the whole population and the abandonment of the island by the Danish crown.

On the Moon, the individual craters in such a chain probably arose from a series of rapidly succeeding strikes from multiple impacts. It is conceivable that a large, comet-like body with a loose structure broke up shortly before the impact through gravitational forces. It is also possible that a group of small bodies with similar trajectories might be responsible for the formation of such chains of craters.

A similar event was observed in 1994, when the nucleus of Comet Shoemaker-Levy was torn apart by Jupiter's gravitational forces, and the fragments fell into the planet's atmosphere, one after the other, over a short period of time.

If high-resolution, detailed images of the Moon's surface near large craters are examined, then numerous short chains of craterlets or crater pits are found. They should probably be ascribed to secondary impacts.

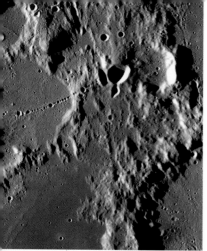

Catena Davy is a chain of small craters.

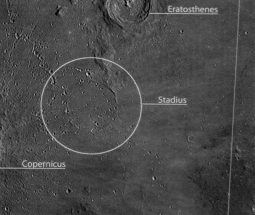

The buried crater Stadius is the best-known example of a ghost crater.

ditatis, which have just two rays like the tail of a comet. The most striking system is that of Tycho, which is even visible with the naked eye. Occasionally such long rays may even reach about 1000 km as, for example, the ray from the crater Glushko (formerly Olbers A) on the Moon's western limb.

Craters with bright ray systems are accepted as being the youngest in the Moon's history. Basically it may be said that the brighter the ray system, the younger the crater. The age of Tycho has been dated to only about 100 million years, and that of Copernicus to around 800 million years.

Smaller craters generally do not show ray systems, but are surrounded by a bright halo. One example of this is Censorinus (3.8 km) on the southeastern edge of Mare Tranquillitatis, which is in fact surrounded by a flat, very bright halo. Under the illumination at Full Moon, the halo of Censorinus is one of the brightest areas of the lunar surface.

Older craters also show the remnants of ray systems. Here, however, the once bright anorthositic rocks have been darkened by lunar erosion.

Ghost craters

The term 'ghost crater' is not one found in the classical lunar nomenclature, but has clearly become established over the course of the last century. The alternative terms 'ghost rings' and 'ghost ring craters' are also sometimes encountered.

Ghost craters are those craters whose crater walls are only detectable under extremely low angles of illumination. They are normal craters, that have either been completely buried by ejecta from large impacts, or have been submerged by lava flows. Examples of craters hidden beneath ejecta are Stadius, which has been buried beneath the secondary rocks from the Copernicus impact, and the crater directly north of Aristillus.

Examples of lava-flooded craters are Lambert R in Mare Imbrium, which is rarely recorded on many old lunar maps, and Wolf T in Mare Nubium.

Craters with ray systems

Under high illumination, many craters exhibit a system of bright, almost radially symmetrical rays. They were created by the impact that formed the crater. Because of the great energy, deeper layers of the bright, anorthositic, lunar crust are uncovered and the material is flung out in rays from the centre of the impact and, under the influence of the weak lunar gravity, falls back onto the darker surface regolith.

These ejecta consist of massive blocks of rock, boulders, and rubble all the way down to completely broken-up material in the form of dust and powder that has the consistency of flour. Such rays are very clearly visible if they cross the dark basalt lava of maria. Ejecta from highland regions is naturally brighter, because the anorthositic material is directly accessible. Segments of the rays consisting of pulverized material have a high albedo (reflectivity), because pulverized material reflects sunlight more strongly than a mixture of larger and smaller particles of rock.

Some craters have clearly asymmetrical ray systems. Experiments by NASA in the 1950s and 1960s showed that they occur when the impact with the Moon's surface occurs at a very shallow angle (< 3°), and is thus almost a grazing impact. The shallower the angle, the more narrowly restricted and asymmetric are the areas covered by the ejected material.

Examples of asymmetrical ray systems are those of the crater Proclus on the western highlands of Mare Crisium, the crater Mädler east of Theophilus (which is fan-shaped), and the craters Messier and Messier A in Mare Fecun-

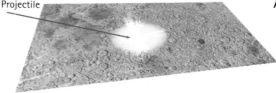

Asymmetrical ray systems are formed by impacting bodies that approach at extremely shallow angles (after Wood).

Unusual lunar features

There are two further classes of surface features on the Moon that are not included in the classical nomenclature.

DMD regions

DMD is the abbreviation for Dark Mantle Deposit. DMD regions are areas of the lunar surface that have been covered in dark volcanic ash, which originates from the mantle, and which may be attributed to strong, eruptive pyroclastic volcanism.

The crater Proculus has a particularly striking asymmetrical ray system.

DMD regions are the observable remains of vigorous active volcanism in 'recent' times, because they largely overlie all earlier changes of the lunar surface that have been caused by impacts and their ejecta. In at least one of the regions (southeast of Copernicus) large instruments reveal lunar domes with summit craters and fractures within a DMD region, and which were the feeder channels for the ash and lava eruptions.

The ash was produced by what is known as pyroclastic volcanism, which is known on Earth on Hawaii or on Iceland (most recently in March and April 2010 from Eyjafjallojökull in Iceland), through fire fountains from chains of vents or fissure eruptions.

The lava rising from deep within the Moon was very fluid and was carrying a large proportion of compressed gas (probably carbon monoxide) under high pressure. When the lava reached the surface through the vent, the gas expanded explosively in fractions of a second and the lava positively exploded, sending it in fountains hundreds of metres – perhaps, being on the Moon, a few kilometres high. This process may perhaps be likened to what happens to a bottle of sparkling water that is opened, when it has been roughly shaken beforehand.

The lavas were rich in iron and titanium and were thus darker in colour than most of the lavas that filled the basins. These particles of fluid lava cooled very rapidly, and crystallized into tiny glass-like spheres and 'rained' down – together with other ash particles – onto the lunar surface. Because of the Moon's weak gravity, the lava probably reached great heights above the surface, and the ashes were spread correspondingly widely around the feeder vents and fissures.

The only samples from a DMD region that have been analyzed in the laboratory came from Apollo 17, and which were gathered near the small crater Ching-Te. This material is in the form of glass spheres and has an unusual dark orange colour, and has therefore become known as 'orange soil'.

Mare Vaporum is an almost circular, basin-like depression directly west of the crater Manilius, and which was first flooded by lava in the Eratosthenian epoch. The fact that the lava plain is relatively young may be judged from the lack of larger craters in the surroundings. Between Manilius and the Hyginus Rille, which is also probably largely of volcanic origin, there lies one of the dark DMD regions. The region itself is part of the Imbrium sculpture, and was thus formed by the Imbrium impact, and much later was covered in dark ash.

Two further DMD regions lie in Sinus Aestuum, one directly east of Copernicus. The Aestuum lava consists of titanium-poor basalt and has been dated to an age of 2.3 billion years, and is thus relatively young. Because the dark DMD ashes do actually cover the Aestuum lava, they must be even younger. Another DMD region, which is crossed by a ray system, lies on the southwestern edge of Mare Serenitatis near the young, 12-km-diameter crater Sulpicus Gallus. Many other lunar domes lie west of Copernicus, some in related groups and a megadome plateau.

Swirls and magcons

The 90-km diameter crater Goddard lies on the northern edge of Mare Marginis. It is unspectacular, except for a small bright area north of it. The bright area is the end of a so-called swirl that has encroached from the lunar farside. Swirls are mysterious features and are associated with anomalies in the average strength of the Moon's magnetic field, which are known by the abbreviation magcon, standing for 'Magnetic Concentration'. The magnetic-field strength over swirl regions is higher than the average magnetic field.

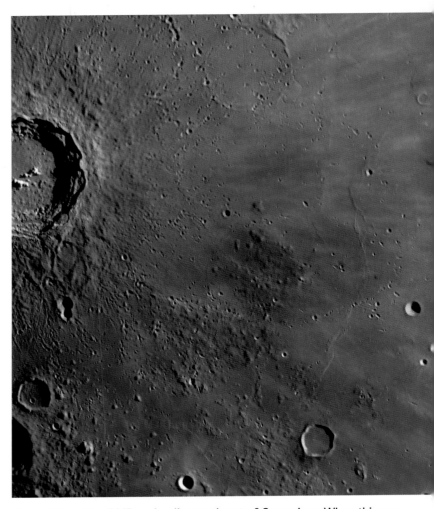

A very interesting DMD region lies southeast of Copernicus. When this area is near the terminator, large telescopes show a few lunar domes with summit craters and fissures, which are undoubtedly the remains of the chain of volcanoes from which the dark ash was ejected.

Swirl regions are rare, apart from the object near Goddard there is just one on the nearside, the Reiner Gamma formation on the northwestern limb of the Moon. All other swirls lie on the lunar farside. Swirls are regions that are covered in material with a very high albedo, and thus are highly reflective of sunlight. The distribution of the material is patchy, irregular, and disordered.

Reiner Gamma lies in Oceanus Procellarum, west of the crater Reiner, and near the landing sites of Luna 8 and Luna 9. It is an elliptical, bright, ring-shaped feature, with a long projection to the northeast. The height difference between the bright material and the dark lava is essentially zero. Even during the Apollo passes over Reiner Gamma, no form of relief could be detected. From the high albedo it may be concluded that the material is pulverized and that the layer is extremely thin.

What is known is that all the swirls – with the exception of Reiner Gamma – lie at the antipodes of the centres of major impact basins. This also explains why there are no additional examples on the nearside of the Moon, because on the farside, with the exception of Mare Orientale and the South Pole-Aitken Basin, there are no large impact basins.

One theory suggests that the swirls might be remnants of impacts of cometary nuclei. But this theory cannot explain why swirls are not distributed more-or-less equally over the whole surface of the Moon – like small craters. Another theory describes the swirl material as bright, pulverized anorthositic dust from the lunar surface, laid down during a normal impact. The stronger magnetic field could have partially repelled electrons and protons from the solar wind so that the material has not yet darkened. But the origin of the magnetic-field anomaly is not explained by this theory.

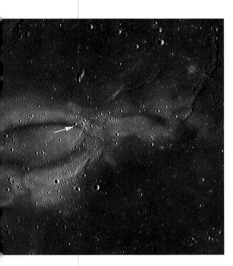

Reiner Gamma is a so-called swirl, where bright material covers the surface.

Lunar geologists are, as yet, unable to conclusively interpret swirls and the associated magnetic anomalies; so far they are features that are simply not understood.

Observation of the Moon

When the Moon is observed with the naked eye it has an apparent size of about 0.5° on the sky, and may be regarded, to a close approximation, as spherical, with a mean diameter of 3474.2 km.

The selenographic coordinate system

To denote the exact location of any point on the lunar surface, we require – exactly as on Earth – two coordinates, namely lunar longitude (λ, lambda) and lunar latitude (β, beta).

The selenographic coordinate system's primary great circle is the Moon's equator, which lies perpendicular to the rotation axis. The rotation axis links the north pole (N) and the south pole (S). The equator and the prime meridian divide the surface of the Moon into four quadrants.

The east-west orientation of the selenographic coordinate system was laid down in 1961 in a decision by the IAU to accord with the astronomical sense – and thus as for an observer on the Moon. To such an observer, the Sun rises in the east and sets in the west. When viewing the Moon with the naked eye or with a non-inverting telescope, from the Earth's northern hemisphere, from the definition east is to the right, and west to the left. All modern lunar charts since 1961 are thus oriented. Earlier Moon maps were published with the astronomical orientation, where the position of east and west corresponded to the orientation on the celestial sphere: west to the right and east to the left.

The maps in this atlas are oriented as if seen through a non-inverting telescope or with the naked eye, namely with north at top and east to the right. (For any variations from this orientation of the images, should the occasion arise, see the appropriate captions or the descriptions in the text.)

Selenographic latitude circles run parallel to the equator. Northern latitudes are expressed as (+) positive numbers, and southern ones as negative (-) ones. It is customary, however, for the abbreviations N (north) and S (south) to be used instead of the mathematical signs. So 60.4°S is identical to -60.4°.

Longitudes (the meridians) run perpendicular to the equator. They are given as positive towards the east and negative towards the west, running from 0° to 180°. A longitude of 180° corresponds to the central meridian on the farside of the Moon. It has become the practice to use directions, as on the sky, for selenographic longitudes, so that 45.3°W is equivalent to λ = -45.3°.

The zero point for the coordinate system lies in the very centre of the Moon's disk at a time when libration in both latitude and longitude is 0° (as observed from Earth), and is in Sinus Medii between the craters Bruce and Oppol-

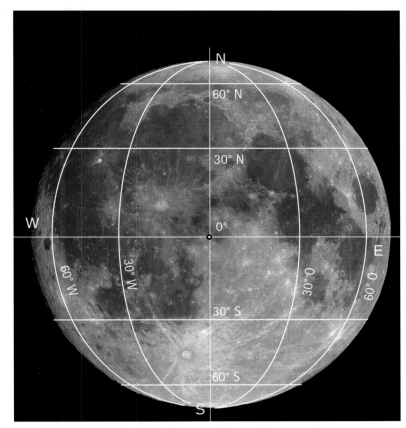

The lunar coordinate system.

zer. The crater closest to the zero point is Bruce, with a diameter of 6.7 km at the coordinates of 1.1°N and 0.4°E.

Mösting A, a bright, sharply defined, round crater with a diameter of 13 km, lies on the western wall of Flammarion and is a basic reference point for the selenographic coordinate system (according to Davies, 1987). Its coordinates are: 3° 12' 43.2"S and 5° 12' 39.6"W.

Height information

For the determination of the position of a point in three dimensions, information about height is also required: the distance of the point from the centre of the Moon or from a defined reference plane. On the Earth's surface, in practice, we use height above a nominal zero plane, that is the height above mean sea level.

The zero point for the lunar coordinate system and the location of the reference crater Mösting A. The spacing of the latitude and longitude coordinate grid is 4°.

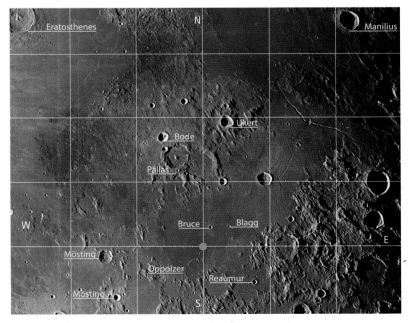

If more precise height information is required (for example, to document changes in elevation over time), the reference plane that is used is the Earth's mean geoid.

The height data given in the descriptions of individual lunar features in the mapping section of this book are defined as follows:
- height data for mountains, mountain ranges and crater walls are generally relative to the surrounding terrain
- height data for central peaks and central mountain ranges in craters and walled plains generally relate to the crater floor
- depth data for fracture zones or rille systems are similarly relative to the average height of the surrounding terrain.

The terminator and colongitude

The term 'terminator' is applied to the boundary between day and night on the Moon, and thus the region where the Sun is actually rising or setting. Because of the lack of a lunar atmosphere, there is practically no scattered light, and thus no twilight zones, so the terminator is a sharp boundary between light and shade. The position of the terminator (ignoring libration in latitude) therefore approximately follows the curvature of the longitude meridians in the lunar coordinate system.

Colongitude is the selenographic longitude on the equator at which the Sun is rising at any given time. The value is given in many planetarium programs and also in printed yearbooks under information about the Moon and is mainly used for planning observations.

Colongitude begins (with waxing Moon) at -90° for the eastern limb of the Moon (Mare Crisium); at Half Moon the colongitude is 0° (the crater Plato in the north, and Maginus in the south are then half illuminated); and corresponds to +90° at Full Moon. Frequently colongitude is only given as a positive angle. The values then correspond to 270° at New Moon, 180° at Half Moon, and 90° at Full Moon.

Libration

Because of the fact that the Moon has captured rotation, an observer on Earth ought to be able to see exactly 50 per cent of the Moon's surface. In reality, things are more complicated and in fact up to 59 per cent of the surface may be observed from Earth, because the captured rotation is subject to certain fluctuations, which are known as libration. Libration is a rocking motion of the Moon in its orbit and consist of three principal components.

Libration in longitude is an apparent rocking of the Moon in the east-west direction. It arises because the Moon rotates at a constant rate around its polar axis, but its velocity continuously changes throughout its elliptical orbit around the Earth. So under favourable libration conditions one may be better able to observe features on the eastern or western limb of the Moon, or see features that

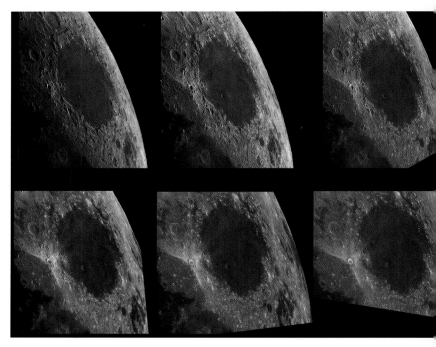

The effects of libration in longitude and latitude on Mare Crisium as an example. The images in the series of photographs (from top left to bottom right) were obtained on the 18th, 19th, 20th, 21st, 23rd, and 24th April 2010. In the last two images, Mare Marginis is visible, whereas in the first three it cannot be seen.

are quite invisible at unfavourable libration angles. The maximum value for libration in longitude can reach ±8°.

Libration in latitude is an apparent rocking in the north-south direction. It occurs because of the inclination of the Moon's equatorial plane relative to the ecliptic. This angle amounts to 6.68° (5.14° + 1.54°). The Moon's rotation axis – like the Earth's rotation axis in its orbit – maintains a fixed position in space. As a result the north and south poles of the Moon alternately incline towards Earth.

Libration in longitude and in latitude are permanently superimposed. The regions of the Moon's surface that are thus rendered visible are know as the libration zones. In addition to libration in longitude and latitude there is also what is known as diurnal libration (caused by parallax). Because of the Moon's relative closeness to the Earth, observers at different points on the Earth see the Moon from slightly different angles. This is of no significance for practical observation.

Apart from apparent libration, it is now known that there is yet another physical libration. It stems from slight changes in the rate of rotation of the Moon about its polar axis. It has no significance for practical observation, because the resulting libration angles amount to just a few arcminutes. Physical libration is constantly monitored by lunar laser ranging.

Because of libration in longitude, in latitude, and diurnal libration more than half the surface of the Moon is visible from Earth.

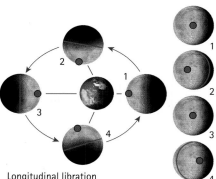

Longitudinal libration

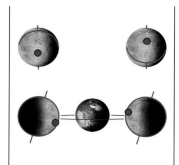

Latitudinal libration

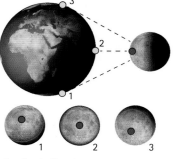

Parallactic libration

Visual observational techniques

Visual observation of the Moon with any size of telescope is fascinating. No other celestial body provides such a three-dimensional perspective, like that of a space traveller, with one's own eyes. A 100-time magnification corresponds to the appearance from a space capsule orbiting the Moon at a height of 3800 km.

Observation of the Moon is worthwhile in any telescope, whether with a large, observatory instrument or with a small, portable refractor.

Telescopes

Because of the plentiful light from Moon even relatively small telescopes are suitable for detailed observations. Even binoculars may be usefully used.

In lunar observation the resolution requirements have other values than those for the separation of double stars, where the Dawes and Raleigh criteria are quoted. Because of the areal extent of features being observed, the diffraction images overlap, and the resolution requirements thus increase significantly. Empirically, resolution values have been determined for

- a black line on a bright background (e.g., for a rille): 23/telescope aperture in mm
- a black point on a bright background (e.g., for a craterlet): 39/telescope aperture in mm

Requirements for resolving lunar features		
Telescope aperture	Rille	Craterlet
60 mm	0.76" ≥ 1.4 km	1.3" ≥ 2.4 km
80 mm	0.58" ≥ 1.0 km	0.98" ≥ 1.8 km
100 mm	0.46" ≥ 0.8 km	0.78" ≥ 1.4 km
150 mm	0.30" ≥ 0.6 km	0.52" ≥ 1.0 km
200 mm	0.24" ≥ 0.44 km	0.40" ≥ 0.8 km
250 mm	0.18" ≥ 0.32 km	0.32" ≥ 0.6 km
300 mm	0.16" ≥ 0.30 km	0.26" ≥ 0.4 km

In reality, these values are considerably reduced by the actual, more difficult contrast ratios as well as by the seeing, and on most night are halved down to 50 per cent. Features with diameters or widths of less than 0.15" are practically unobservable from the Earth's surface, even with large telescopes. The actual resolution ability may be determined from special charts. The conversion factors for the sizes on the lunar surface are, however, only valid for the centre of the Moon, not its limbs.

The theoretically achievable values are restricted in reflecting telescopes by the secondary mirror in the centre of the light path. The contrast performance is defined by the difference between the diameters of the primary and secondary mirrors: a 200-mm reflector with a 60-mm secondary thus delivers the contrast obtained by a 140-mm instrument without any obstruction.

Magnification and seeing

For studies of individual lunar craters a magnification of at least 100× needs to be obtained. To be able to use the higher resolution, magnifications that are significantly greater than the resolving magnification of (telescope aperture in mm)/0.7 are recommended. Frequently, under good conditions it is possible to observe near the maximum magnification of (telescope aperture in mm)/0.35.

However, seeing has an enormous influence on the magnification that can actually be reached. Because the moving atmospheric cells frequently have a diameter of between 10 and 15 cm, telescope apertures in this range give their best performance. On most nights larger telescopes do not deliver better images, and only reach their potential on the few calm nights. With smaller telescopes, on the other hand, the seeing rarely sets a limit on performance.

In practice, for telescopes with apertures of 50 to 120 cm, working magnifications provide nearly the theoretical maximum magnification. Larger apertures should, however, be used with at least the resolving magnification, so that the resolution is not reduced.

Photographic observational techniques

It has become very easy nowadays for the amateur to obtain high-quality images of the Moon. Over and done with is the time of chemical photography, with its long exposure times – essential because of the long focal lengths – and the subsequent work in the darkroom. There is a gulf between modern-day digital imaging results and chemical photography.

In the 1980s, professional astronomers introduced the method of 'lucky imaging'. In this, video cameras (such as webcams or surveillance cameras) were introduced, which, with their rapid succession of images and short exposure times took hundreds or thousands of images. Because of the short exposure time, the individual images were very faint, but in moments of best seeing, they were absolutely sharp (hence 'lucky' imaging). Subsequently, the best individual images were identified on the video film, stacked on top of one

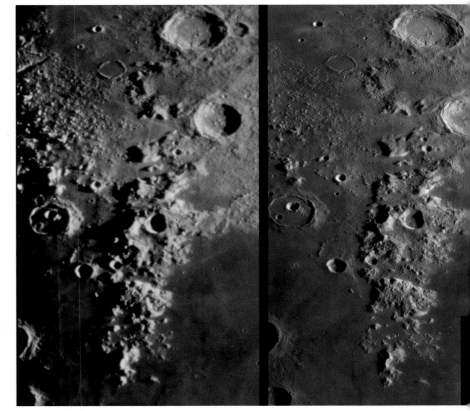

Lunar photography has seen major advances in recent years. These two images show the enormous increase in quality and sharpness between modern digital techniques (right) when compared with old analogue photography (left).

Charts for testing resolving power

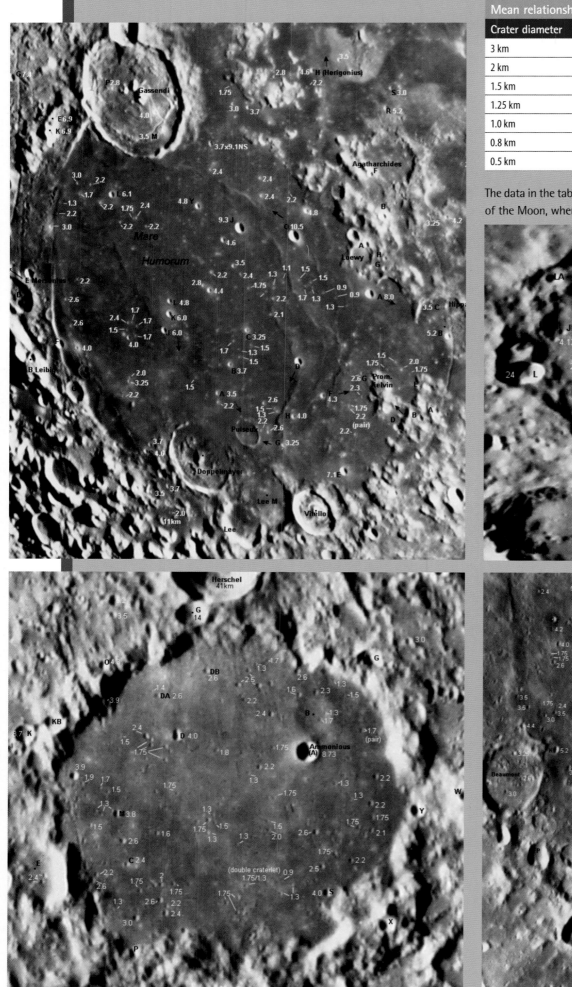

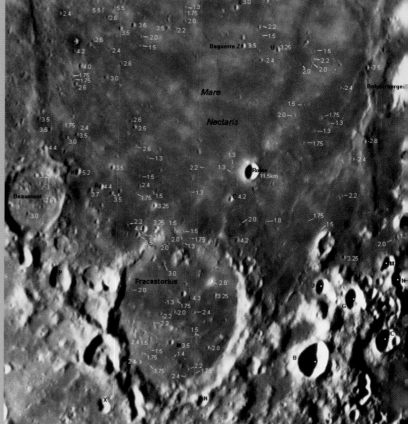

Mean relationships between size on the Moon and angular size

Crater diameter	Angular diameter
3 km	1.61"
2 km	1.08"
1.5 km	0.81"
1.25 km	0.67"
1.0 km	0.54"
0.8 km	0.43"
0.5 km	0.27"

The data in the table correspond to those at average distance of the Moon, where 1" corresponds to 86 km.

The resolution charts are from the Lunar Orbiter Atlas, after M. Clark. The values are in km, and the accuracy corresponds to about 100 m.

another and added together. The addition reduced the electronic noise. At the same time, very effective routines for sharpening images (e.g., wavelet filtering) were programmed.

Mounts

Great demands are not placed on the mounts. Tracking on the right-ascension axis should be available. An adjustable speed to track the Moon is not absolutely essential, but an advantage. If it is not available, because of the variation in its velocity during the sequence of images, the Moon slowly drifts across the imaging camera's chip, which is normally extremely small, and the software later crops the edges of the frames, because differing surface features are imaged between the first and last images of the sequence, which cannot be brought into alignment.

Even when the tracking of the mount has the same speed as the movement of the Moon, the Moon still drifts slowly across the chip with long image exposure times, because of its change in position in declination. If the mount is a high quality one, then the position of the image of the Moon may be counteracted by hand, using the slow-motions. These corrections should be made gently, and the image of the Moon should not be shifted in jerks or backwards.

The mount must be sufficiently stable for the size of instrument it carries. For one thing, it should not – when set up outside – be affected by gusts of wind and, for another, an undersized mount will not allow the image of the Moon to be focussed properly.

Telescope

In principle, any optical system is suitable for lunar photography. In each one, however, different factors need to be taken into account. Regardless of whether it is a refractor or reflector, the instrument must provide a contrasty image free from scattered light, otherwise the optics will not be able to image the smallest craters or faint details.

The same applies to the collimation of the imaging optics. It should be adjusted as perfectly as possible. The collimation of refractors is extremely stable. For Newtonian and Schmidt-Cassegrain telescopes, especially when used as portable instruments, on any evening when a series of images is planned, the collimation should be checked shortly before the start. The optics should, of course, be manufactured as perfectly and precisely as possible.

Any optical system must be allowed to stabilize before imaging begins, which means that the temperature of all the system components, such as the tube and objective or mirror must have reached the surrounding air temperature. With reflecting telescopes, the process – depending on the diameter of the optics – takes a good two hours. The relatively large masses of glass in expensive triplet refractor objectives also require time. Open reflecting telescopes, such as Newtonians, are prone to problems of tube currents, with a continuous exchange between the air in the 'tube' and the surrounding air. Larger instruments require significantly more time until they become thermally stable.

For high-resolution lunar photography, the aperture of the optics is decisive. The larger it is, the higher is the angular resolution on the Moon. The 'lucky imaging' method yields image results that attain the theoretical resolution of particular optics. The larger the diameter of the imaging optics, however, the more it suffers from seeing effects. Very good image results may be obtained with apertures of just 100 to 150 mm.

Long focal lengths are required, which is why Schmidt-Cassegrain telescopes or older, classical, long-focal-length refractors, like those found in many public observatories, are employed. Short focal lengths must be extended with additional optical elements. Normal eyepieces are calculated to give parallel rays from the exit pupil and are thus not very suitable for the standard eyepiece projection method. Barlow lenses or teleconverters are better (for example, the fluorite Flatfield Converter by Baader, which employs intermediate rings to give extensions of the focal length between 2× and 8×).

Imaging techniques

For imaging with CCD-sensors there is an optimum ratio between the image scale (focal length) and the size of the pixels on the imaging sensor. If the imaging focal length does not conform to the pixel size, then the resolution is 'wasted'. Few amateurs enjoy constant and perfect seeing at their observing location, however, that would enable them to adjust their focal length (requiring larger apertures) to the pixel size. So the aperture or the focal length (or both) that have to be adjusted to suit the seeing conditions.

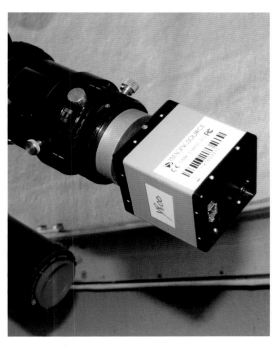

For lunar photography, digital cameras such as model DMK31 from The Imaging Source are very suitable.

Focussing of the image of the Moon may prove difficult. The position of the image focus continuously changes between sharp and unsharp. As a result there are geometric changes in the image. Seeing effects are significantly greater both inside and outside the exact focal point.

To begin with, the best possible setting is found by rapidly changing between inside and outside the point of focus (seeking optimum image sharpness and image contrast). Then a small region with as many craterlets as possible is chosen and the image is observed on the monitor for some seconds. Now the focus setting is fractionally altered (either moving inside or outside the focus) and the image again examined for a few seconds. Finally, the focal point is shifted slightly in the other direction. During these few seconds, the seeing conditions will always be optimum for a few tenths of a second, so the optimum position of the focus may be found with some certainty.

With a Schmidt-Cassegrain system, where focussing is normally carried out by shifting the primary mirror, the use of a Crayford or helical focussing mount is preferable for this task. When the primary mirror is shifted, it may tip slightly and ruin the careful collimation that has been carried out. For this reason, focussing with the mirror should always be carried out in the same direction, both in the adjustment of the optics and in focussing the image of the Moon. The final focus may then be obtained by using the Crayford or helical focussing mount.

If the seeing conditions are very variable and not particularly good overall, then it may be advisable to turn the electronic gain of the camera to its maximum setting. The images obtained are then very faint, but the exposure time for the individual images can be drastically reduced, which helps to 'freeze' the seeing.

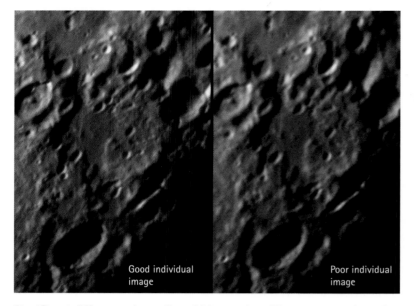

Good individual image

Poor individual image

Significant differences in quality within a series of images occur through different seeing conditions.

With focal ratios of f/20 to f/30 there is a great tolerance in focussing, which may actually be millimetres in extent. That means that if one alters the focal plane of the camera with the eyepiece movement of 1 or 2 mm, the sharpness of the image hardly changes. For this reason, manual focussing is recommended for obtaining sharp lunar images, because the focussing tolerance may be covered more rapidly. Electrical focussing, which is carried out by stepping or servo motors as a rule change the focus of the camera only in steps of a tenth or a hundredth of a millimetre.

Once it is definite that the best possible focus has been achieved, then the imaging sequence may begin. If individual frames are obtained over a long period of time, then it is essential that the focus is checked at ever shorter intervals, because it may alter as a result of temperature changes.

With poor seeing the imaging frequency should be high, such as 30 images per second, to capture as many individual images as possible during the few moments of good seeing. The exposure times of the individual images should likewise be as short as possible (< 1/30 s). This results in relatively faint individual images, and the total number of individual images is correspondingly high (2000 to 3000 images). With good seeing the imaging frequency may be reduced and the exposure times lengthened. The indivi-

The use of colour filters can be of advantage, particularly when photographing with achromatic refractors.

dual images will be stronger, and so the total number of images may also be reduced (about 750 to 1250).

Imaging of features near the terminator, especially those that are relatively low in contrast, such as mare wrinkle ridges, lunar domes, etc., require a higher number of stacked individual images, to be able to produce a completely noise-free final image, in that region between light and shade.

The two values gain (the degree of electronic image amplification) and gamma (the contrast between black and white) as well as exposure times should be balanced against one another so that the full range of the histogram (that is, the distribution of pixels between black and white) should be completely utilized. In evenly illuminated regions of the Moon, the gamma may be set high, whereas regions near the terminator should be obtained with lesser image contrast.

If seeing conditions are optimal and very good results may be expected, it is advisable to take multiple images of the same region of the Moon with differing settings of exposure, contrast and amplification, and process them later to give the best result in the final image.

Filters

A good optical, plane-parallel, UV/IR blocking filter, which removes the spectrum below 400 nm and above 700 nm – as well as protecting the chip from dust particles – should always be present in the optical train. In lunar images in particular, dust particles make themselves very unpleasantly obvious.

With all (even good) refractors, and also in all optical trains in which a focal extender with refractive optics (such as a Barlow lens, etc.) is used, an additional very weak yellow filter is advisable, because then the critical blue spectral region will be removed. With cheap refractors it is advisable to insert a green bandpass filter (in combination with the UV/IR blocking filter), even if the image brightness decreases and the exposure time increases. The filter removes from the visible spectrum precisely the range for which refractors are optimized.

Red filters – such as the Schott RG 610 or RG 645 – reduce the effects of turbulence, because seeing effects are less pronounced at long wavelengths than in the short-wave, blue region of the spectrum. So-called infrared pass filters work even better, and which are only transparent between about 700 nm to 750 nm. When used, the UV/IR blocking filter must be removed.

Filtration has some disadvantages, however. First, it reduces the image brightness and the exposure time must therefore be increased. Then it reduces the resolution, the farther into the red that one works, because the resolution for any optical system – whether a reflector or refractor – depends on the wavelength and is highest in the blue. If seeing conditions permit, a blue bandpass filter should be employed (removing the green, yellow, and red spectral regions) for the best possible results.

Image processing

After the image sequence has been stored, the video images are imported into appropriate processing software (such as Avistack, Registax or Giotto) as individual frames. In the first working step, any displacements – caused by seeing effects or inadequate guiding (or both) – are determined for all the images relative to a reference image, and any deviations are corrected, so that all the images may be exactly superimposed. To do this a series of reference points is determined. This step in the process is known as registration or registering.

In the following step, areas of the images are discarded in which there is no point in setting reference points for later processing, which will therefore be overexposed regions and – if present – the dark image background (or both). Then follows establishment of the reference points. They will be required later, so that the best sections of each of the individual images may be superimposed exactly and then co-added. Subsequently, every individual image of the video sequence is divided into square sections, of variable size, and the image quality is analyzed for every individual image and every section.

This is followed by the actual image processing using the reference points. The individual image sections are now compared with one another at the sub-pixel level. This stage is the most time-consuming and is dependent on the speed of the computer, its internal memory capacity, and the number of reference points. When processing of the raw video sequence has come to an end, the best (the sharpest and most contrasty) image sections from each individual frame are co-added to give a final image, so that every image section consists of the same number of units as the preset total number of individual images that are to be co-added. This step is known as stacking and when it has ended the finished raw image is stored and it ready for further processing.

Most of the images by W. Paech published here consist of about 150 individual images, the majority extracted from series of between 1000 and 2000 raw images. The stacking rate is therefore between 7.5 and 15 per cent. Under difficult imaging situations, in regions between light and shadow (e.g., mountain shadows on crater floors), the stacking rate has been increased to as much as 20 per cent to minimize noise in the darker regions.

Once the co-added raw image has been created, sharpening follows, with either wavelet or Mexican Hat, or a similar filter function. Eventually, final image processing is carried out with normal image processing software (such as Photoshop or a similar program), which generally just consists of adjusting the grey scale of the histogram. If the raw co-added image is of good quality, the stacked raw image has been sharpened, and the chosen region of the Moon more-or-less evenly illuminated, there is actually little need for final image processing to obtain the finished result. If the raw images are slightly oversharpened, then this shows as slight noise in the darker portions of the image, and very conspicuously in uniform mare areas. A slight softening may then be applied with a blurring filter.

Things are different if the chosen region of the Moon is near the terminator, and has been imaged with relatively short focal lengths of two to three metres and a chip size of 1024 × 768 pixels or more. Then the field of view is relatively large and the camera has to form an image with an enormous difference in brightness between light and dark. However, the CCD or CMOS chip in a video camera can only record the range from black to white as a certain number of steps of grey. If, for example, a group of lunar domes and their summit craters or a mare wrinkle ridge is to be photographed and, because of their restricted height, need to be imaged near the terminator, then exposure times result that cause the image of craters that are already fully illuminated by the Sun to become saturated. Saturation means that in the white-tinted areas, no steps – and thus no details – are visible. Any subsequent image sharpening increases the contrast between light and dark even more, and the resulting images lose all aesthetic appeal. So if relatively dark features near the terminator must be imaged in detail at the same time as bright crater rims, the same area of the Moon's surface must be imaged with two different exposure times, each appropriately adjusted for brightness. The finished final images from each series are then superimposed and mixed. In image processing this technique is known as using layers.

When superimposing two or more images with differing exposure times, standard image-processing programs (such as Photoshop or similar pro-

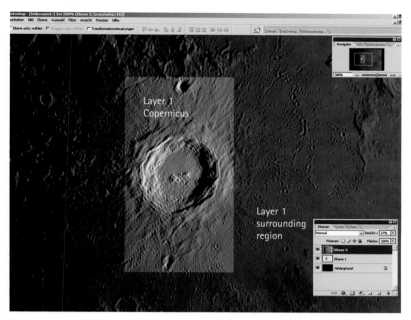

An example of the use of layers. The first (background) layer is a short-exposure section of an image with the crater Copernicus, where the bright crater walls still show structure. The superimposed, second layer is an exposure with a longer exposure of the corresponding region of the Moon, in which the exposure time of the raw images was adjusted to suit the overall surface of the Moon. Both images were perfectly superimposed. Then the second layer was set to semi-transparency, so that the lower layer was still simultaneously visible.

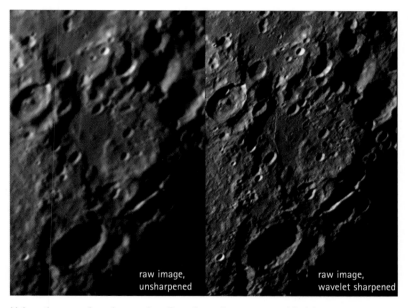

Using the wavelet-sharpening function, the raw co-added image may be improved. The example shows the results before and after use of this filter.

grams) provide various possibilities for working with layers, ranging from using an eraser tool on selected waviness or post-exposure flaws to differential adjustment of graduations with weak selected gradients.

Good – still unsharpened – raw images should always be archived in 16-bit format as uncompressed tif or fit files, so that, should the occasion arise, they may be subjected to further processing at a later date. The brightness and contrast settings of the monitor on which lunar images are processed are very important. Amateurs rarely have access to a monitor suitable for taking things to the pre-print stage. The depiction of images of the Moon on different adjusted monitors is highly critical. The image may appear fine on one monitor, but may appear far too dark and contrasty, or featureless and too bright on another. The actual monitor used should be set correctly using a grey wedge. The background lighting of the room in which the monitor is used also has an effect on the appearance of the image. So the images should be processed under lighting conditions that are as consistent as possible.

Mare Crisium

Neper

Dublago

3

4

Hargreaves

Maclaurin

Mare Fecunditatis

Langrenus

Mare Smythii
2.0°N, 87.0°E

Mare Smythii, 'Smyth Sea', named after a prominent, 19-century amateur observer, is a circular mare on the Moon's eastern limb and appears, because of limb foreshortening, to be highly elliptical. It covers an area of just about 105 000 km², and a significant portion lies on the Moon's farside. Successful observation is only possible under favourable libration angles, but it then presents some interesting features. The outer wall of the basin rises above the lunar horizon, and various lava flows are visible inside it. In the north and south, in particular, there are unusually dark lava surfaces, which, in the south, are associated with the 55 km-diameter, FFC-crater Kiess. Mare Smythii is one of the youngest impact basins and its average depth is about 5 km below the general level of the surrounding surface.

Mare Marginis
12.0°N, 88.0°E

Mare Marginis, 'Marginal Sea', is an irregularly shaped lava area, 360 km across (about 62 000 km²). The mare has a north-south orientation and extends onto the Moon's farside.

Mare Marginis lies to the north of Mare Smythii. On the border between the mare lie the two craters Neper and Jansky. Neper (8.8°N, 84.5°E) is a large crater, 137 km in diameter. The floor of the crater is flooded by very dark lava and there is a visible central peak. Jansky (8.5°N, 89.5°E) has a diameter of 73 km. Both craters are visible under favourable libration conditions.

Kästner

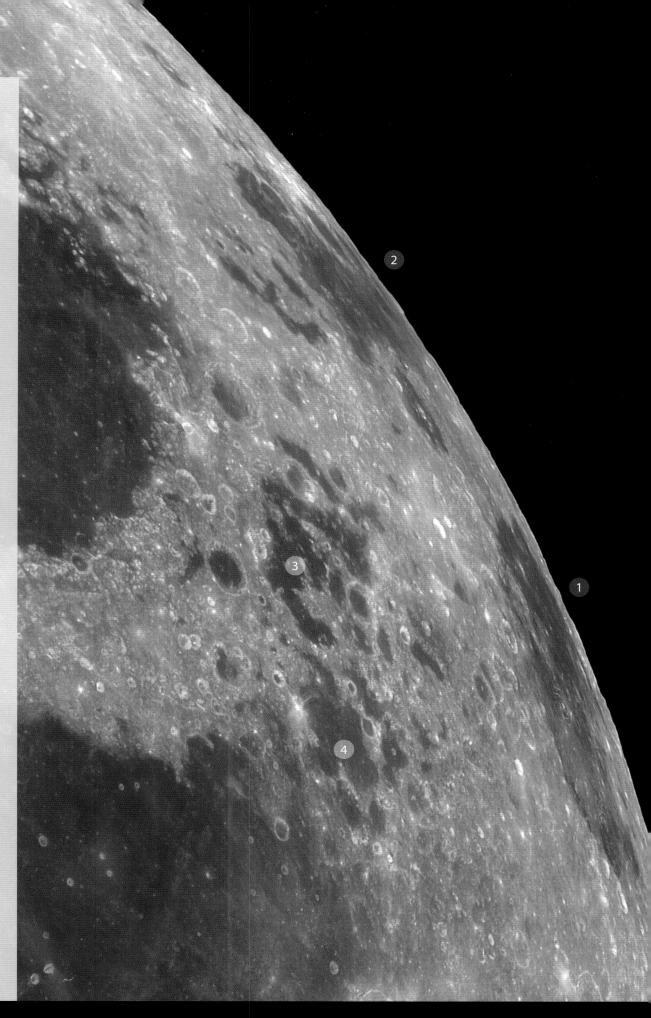

Mare Undarum ③
7.0°N, 69.0°E

Mare Undarum, 'Sea of Waves', is distinctly better placed for observation from Earth. It is a small (only 21 000 km²) area consisting of several eroded craters flooded by lava. The individual named craters in this mare are the small crater Boethius (10 km, 5.6°N, 72.3°E), and the significantly larger crater Dubiago (4.4°N, 70.0°E). Dubiago is 51 km in diameter. The floor of the crater has been flooded by very dark lava, and the lava surface is crossed by several mountain ridges.

Mare Spumans ④
1.0°N, 65.0°E

Mare Spumans, 'Foaming Sea', lies to the south of Mare Undarum and is more favourably placed for observation, being farther from the lunar limb. With an area of about 16 000 km² it is slightly smaller than Mare Undarum and, as far as shape is concerned, is approximately circular. At the southern end lie the craters Hargreaves (16 km, 2.2°S, 64.0°E) and Maclaurin (50 km, 4.9°S, 68.0°E). The latter has a flattened central peak, oriented north-south. To the southwest lies Mare Fecunditatis and even farther south, the large crater Langrenus. The origins of the four maria have been dated to the period 4.5–3.85 billion years ago.

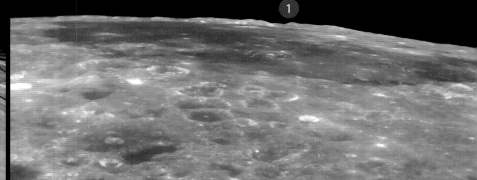

Kiess

Kästner

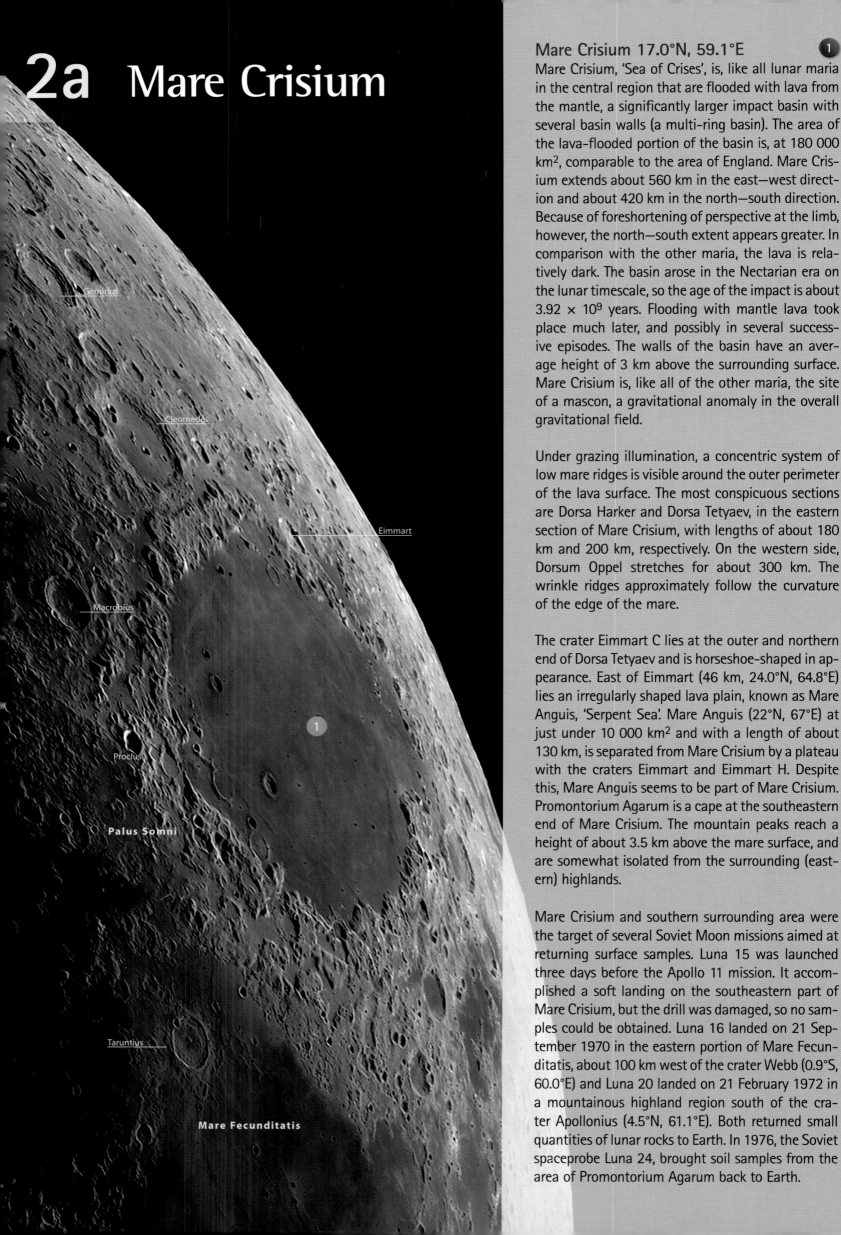

Geminus

Cleomedes

Eimmart

Macrobius

1

Proclus

Palus Somni

Taruntius

Mare Fecunditatis

2a Mare Crisium

Mare Crisium 17.0°N, 59.1°E ❶

Mare Crisium, 'Sea of Crises', is, like all lunar maria in the central region that are flooded with lava from the mantle, a significantly larger impact basin with several basin walls (a multi-ring basin). The area of the lava-flooded portion of the basin is, at 180 000 km², comparable to the area of England. Mare Crisium extends about 560 km in the east—west direction and about 420 km in the north—south direction. Because of foreshortening of perspective at the limb, however, the north—south extent appears greater. In comparison with the other maria, the lava is relatively dark. The basin arose in the Nectarian era on the lunar timescale, so the age of the impact is about 3.92×10^9 years. Flooding with mantle lava took place much later, and possibly in several successive episodes. The walls of the basin have an average height of 3 km above the surrounding surface. Mare Crisium is, like all of the other maria, the site of a mascon, a gravitational anomaly in the overall gravitational field.

Under grazing illumination, a concentric system of low mare ridges is visible around the outer perimeter of the lava surface. The most conspicuous sections are Dorsa Harker and Dorsa Tetyaev, in the eastern section of Mare Crisium, with lengths of about 180 km and 200 km, respectively. On the western side, Dorsum Oppel stretches for about 300 km. The wrinkle ridges approximately follow the curvature of the edge of the mare.

The crater Eimmart C lies at the outer and northern end of Dorsa Tetyaev and is horseshoe-shaped in appearance. East of Eimmart (46 km, 24.0°N, 64.8°E) lies an irregularly shaped lava plain, known as Mare Anguis, 'Serpent Sea'. Mare Anguis (22°N, 67°E) at just under 10 000 km² and with a length of about 130 km, is separated from Mare Crisium by a plateau with the craters Eimmart and Eimmart H. Despite this, Mare Anguis seems to be part of Mare Crisium. Promontorium Agarum is a cape at the southeastern end of Mare Crisium. The mountain peaks reach a height of about 3.5 km above the mare surface, and are somewhat isolated from the surrounding (eastern) highlands.

Mare Crisium and southern surrounding area were the target of several Soviet Moon missions aimed at returning surface samples. Luna 15 was launched three days before the Apollo 11 mission. It accomplished a soft landing on the southeastern part of Mare Crisium, but the drill was damaged, so no samples could be obtained. Luna 16 landed on 21 September 1970 in the eastern portion of Mare Fecunditatis, about 100 km west of the crater Webb (0.9°S, 60.0°E) and Luna 20 landed on 21 February 1972 in a mountainous highland region south of the crater Apollonius (4.5°N, 61.1°E). Both returned small quantities of lunar rocks to Earth. In 1976, the Soviet spaceprobe Luna 24, brought soil samples from the area of Promontorium Agarum back to Earth.

The rings around the Mare Crisium basin

Mare ridges	375 km
Edge of the mare	500 km
Cleomedes ring	635 km
Geminus ring	1075 km

The outer walls of the Crisium basin are not well-preserved, but are still detectable in a few places. Ring 1 is indicated by the mare ridges; Ring 2 forms the border of the mare and Ring 3 runs through the crater Cleomedes (and is broken by the crater). Ring 4 is only weakly detectable and is interrupted by the craters Geminus and Berosus.

Picard 14.6°N, 54.7°E ②
Peirce 18.3°N, 53.5°E ③

Picard and Pierce, with diameters of 22 km and 18 km, respectively, are the only notable craters on the floor of Mare Crisium. Both exhibit sharply defined crater walls and are certainly relatively young when compared with the craters in the southern highlands. Picard is one of the few double-walled craters, that may be observed with small telescopes.

O'Neill's Bridge 15.2°N, 49.2°E ④

O'Neill's Bridge is not an official lunar designation, but a sort of nickname, first introduced by the amateur John O'Neill in 1953. The supposed bridge is visible shortly after Full Moon, when the terminator reaches the western edge of Mare Crisium. When seeing conditions are relatively poor (or the telescope's aperture is too small), the object resembles a bridge, joining the tips of the capes Promontorium Lavinium and Promontorium Olivium. (Neither designation is used officially nowadays.) If the seeing conditions are good and the instrument large enough, two small, eroded crater pits are revealed at the site. O'Neill's Bridge lies directly southeast of the crater Proclus.

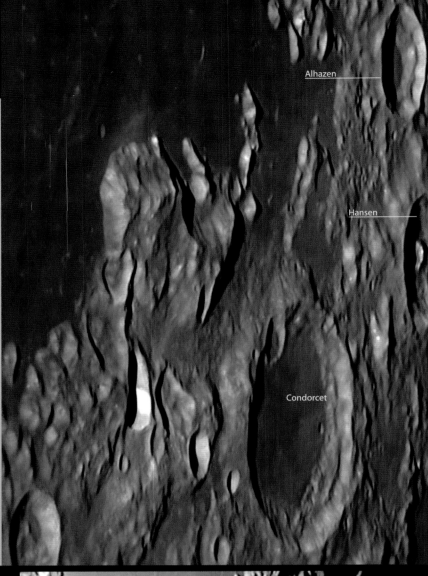

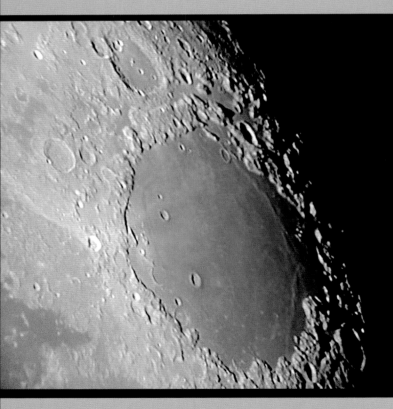

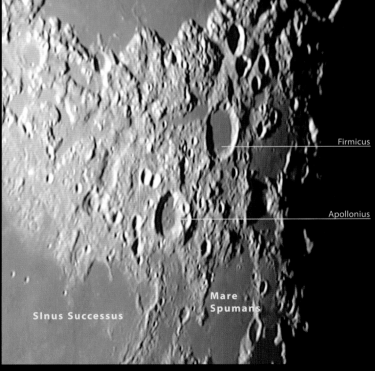

Lick 12.4°N, 54.7°E ⑤
Yerkes 4.6°N, 51.7°E ⑥

Two, structurally almost identical, lava-flooded craters lie southeast and east of O'Neill's Bridge. Lick has a diameter of 31 km, and Yerkes is somewhat larger at 36 km. North of it lies Yerkes E (10 km), which is linked to Yerkes by a ridge.

Dorsum Oppel 18.7°N, 52.6°E ⑦

With a length of about 300 km, Dorsum Oppel is the most noticeable segment of mare wrinkle ridge in the western part of Mare Crisium. Together with Dorsa Harker and Dorsa Tetyaev on the eastern side of Mare Crisium, it forms a nearly complete, concentric ring, following the edge of the mare.

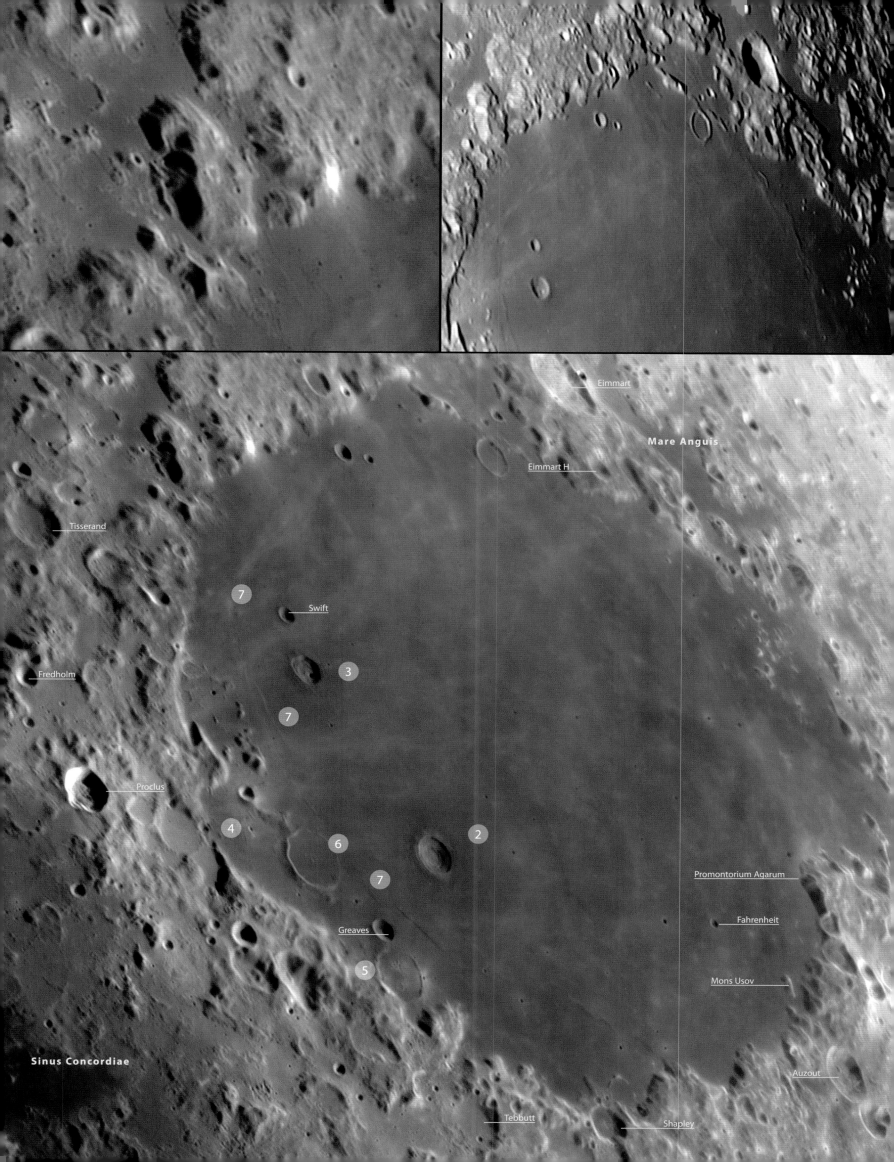

Eimmart

Mare Anguis

Eimmart H

Tisserand

7

Swift

Fredholm

3

7

Proclus

4

6

2

Promontorium Agarum

7

Fahrenheit

Greaves

Mons Usov

5

Sinus Concordiae

Auzout

Tebbutt

Shapley

Cleomedes 27.7°N, 56.0°E ①

Cleomedes is a very conspicuous crater with a diameter of 126 km. There are small craters and a central peak visible on the floor. Across the northern portion runs Rima Cleomedes, a 30 km-long rille that is difficult to observe. (A large telescope, the correct lighting and a favourable libration are required.) The northern wall of the crater has eroded and has been breached by the craters Trailes (43 km, 28.4°N, 52.8°E), Cleomedes A (14 km) and Cleomedes E (20 km).

Geminus 34.5°N, 56.7°E ②
Messala 39.2°N, 60.5°E ③
Burckhardt 31.1°N, 56.5°E ④

North of Cleomedes lies Burckhardt, a 56 km-diameter crater. Burckhardt is superimposed on two smaller craters, Burckhardt E (39 km) and Burckhardt F (43 km). Normally, large craters are superimposed by smaller ones, but with Burckhardt the direct opposite is the case.

Even farther north lie the two craters Geminus and Messala. Geminus with a diameter of 85 km is another striking crater in this area of the Moon. The crater has a central peak and the southern wall is breached by a wide valley. Because of its size of 125 km, Messala would once have been classed as a 'walled plain'. The floor is relatively smooth, but the view in large telescopes reveal some irregularities. Geminus C (16 km) and Messala G (29 km) are smaller craters that lie close to conspicuous rays of impact material, which are visible under high angles of illumination. The origin of the ray system cannot, however, be determined.

Lacus Spei 43.0°N, 65.0°E

5

Lacus Spei, 'Lake of Hope', is a small lava flow, 80 km long, lying northeast of Messala. It lies directly between the craters Schumacher (61 km, 42.4°N, 60.7°N) and Zeno (65 km, 45.2°N, 72.9°E).

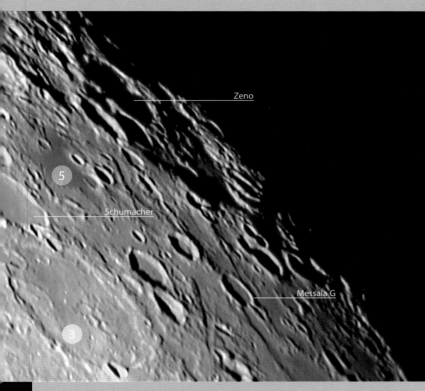

Gauss 35.7°N, 79.0°E

8

Gauss is an extensive, gigantic complex crater with a diameter of 177 km. On the southeastern floor, a series of small craters and hills are visible under favourable lighting conditions. The largest crater on the northeastern rim is Gauss B (37 km), and smaller craters are Gauss A (18 km) and Gauss E (29 km). Because of its extreme limb location, successful observation requires a very favourable libration.

Hahn 31.3°N, 73.6°E

6

Hahn is a medium-sized crater with a diameter of 85 km and a large central peak. The northern wall of the crater is broken by a smaller crater. To the east, outside the crater's rim lies Hahn B (15 km).

Berosus 33.5°N, 69.9°E

7

Berosus is a lava-flooded crater, 74 km in diameter. The crater floor is smooth and flat and shows no detail in amateur-sized instruments. The western wall has been breached by Berosus A (12 km).

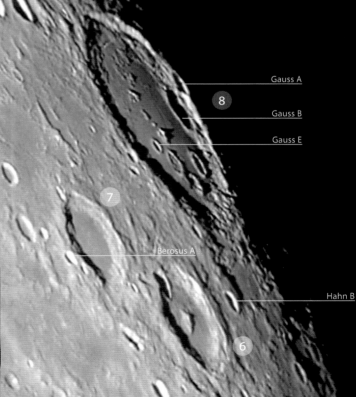

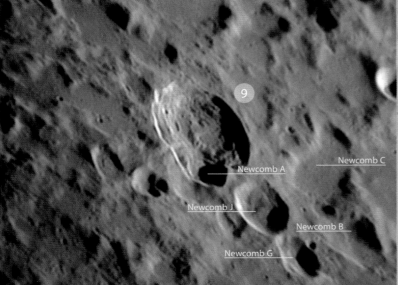

Newcomb 29.9°N, 43.8°E

9

This crater has a diameter of 39 km and a fractured crater floor. Newcomb A (20 km) is superimposed on the southern rim. To the southeast, the craters Newcomb J and C (29 km and 23 km) lie next to one another, and below them, Newcomb G and B (16 km and 23 km), similarly close together. Newcomb C and B are largely lava-flooded.

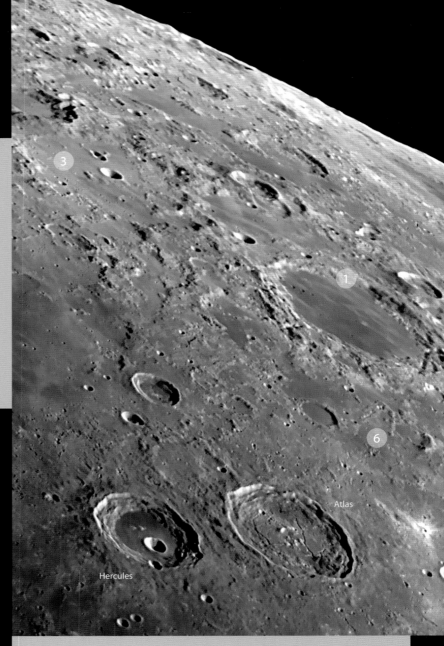

Endymion 53.9°N, 57.0°E ①

With a favourable libration, Endymion is a very conspicuous complex crater on the Moon's northern limb. It has a diameter of 123 km and is 2.8 km deep. The crater's floor is smooth and has been flooded with unusually dark lava; there have been landslides in several places along the walls. The crater was formed during the Nectarian period. Large apertures reveal a few small and minute craters. Three small craters that are almost the same size lie directly to the west of the inner crater wall. Under high solar illumination, a system of bright rays of ejecta, of unknown origin, becomes visible crossing the crater.

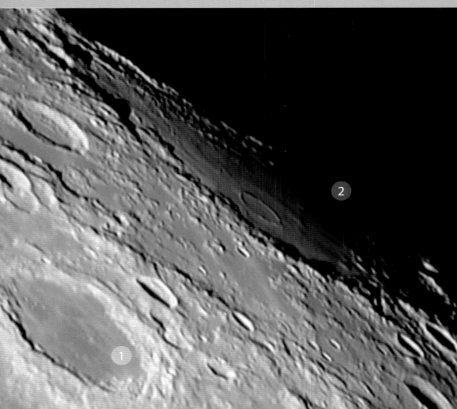

Mare Humboldtianum 56.8°N, 81.5°E ②

Mare Humboldtianum, 'Humboldt Sea', is a difficult object for terrestrial observers, because it lies right on the Moon's northeastern limb. Part of it stretches onto the lunar farside, and as a result, favourable observational conditions are very closely linked to the libration. Mare Humboldtianum is the central, lava-flooded area of an impact basin with several outer rings (i.e., a multi-ring basin), with an overall diameter of 640 km. The lava-covered surface has a diameter of about 270 km. With a favourable libration, larger telescopes clearly show individual structures within the Mare, and under high angles of illumination parts of the lava flow are crossed by bright rays.

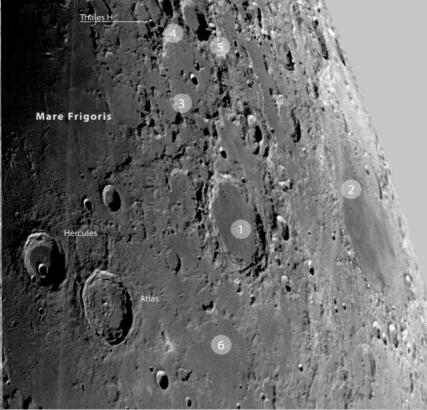

Thales H

Mare Frigoris

Hercules

Atlas

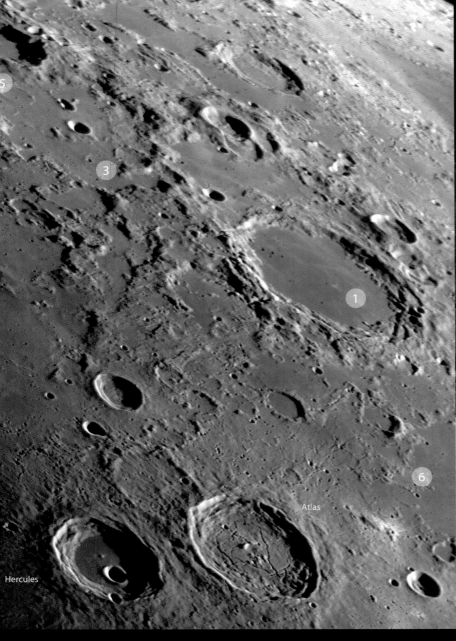

de la Rue 59.1°N, 52.3°E

A large, heavily destroyed walled-plain with a diameter of 136 km, and both large and small craters on the floor. The crater is named after Warren de la Rue, a British amateur astronomer (1815–1889), who, amongst other things, was a pioneer of lunar photography.

Thales 61.8°N, 50.3°E
Strabo 61.9°N, 54.3°E

Thales is a young crater with a diameter of 31 km, circular in shape and with a sharp rim. A small crater, Thales H (10 km) lies outside the southwestern wall. Under Full-Moon illumination, the crater is the source of a bright ray system. Directly east of Thales is Strabo, with terraced inner crater walls and a diameter of 55 km. Strabo – unlike Thales – exhibits no sign of a ray system.

Lacus Temporis 46.0°N, 57.0°E

Lacus Temporis, 'Lake of Time', is a long, extended, lava plain, lying east and southeast of the crater Atlas. The diameter amounts to about 250 km, but because of perspective effects through limb foreshortening, it appears elongated to terrestrial observers.

Atlas 46.7°N, 44.4°E ①

The crater Atlas, with a diameter of 87 km, is a conspicuous example of the class of Fractured Floor Craters (FFC). This is shown by an extensive system of rilles on the floor of the crater. Posidonius and Gassendi are other striking examples.

Atlas has strongly terraced, inner crater walls, which rise 3 km above the crater's floor. The tip of a central peak is visible and the floor of the crater is rough and fissured. The rille system, Rima Atlas, is of volcanic origin (as will all craters of this type), and under high illumination, at least two dark spots are visible on the crater floor (as in the crater Alphonsus). The dark spots probably consist of pyroclastic ash deposits and suggest volcanic activity occurred a long time after the actual impact.

Hercules
46.7°N, 39.1°E

Hercules has a diameter of 69 km and, like Atlas, displays terraced inner crater walls. In contrast to Atlas, however, the crater's floor is flat, smooth and shows – just visible – the tip of a central peak. In addition, the floor of the crater has been flooded by lava. Directly south of the central peak is Hercules G (13 km). A further small crater, Hercules E (9 km) breaches the southern crater wall. Atlas and Hercules form a striking pair of craters on the Moon's northeastern limb.

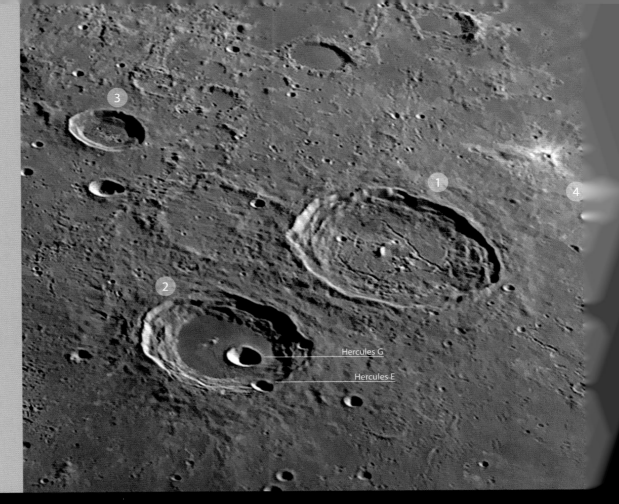

Hercules G

Hercules E

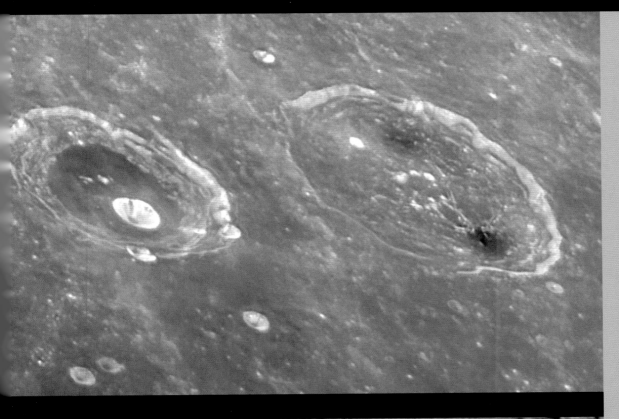

Keldysh
51.2°N, 43.6°E

A younger crater, 33 km in diameter, and with a sharp crater wall. A tiny crater is visible on the southern side of the floor.

Atlas A
45.3°N, 49.6°E

Atlas A is a smaller crater, about 22 km in diameter. The wall of the crater is broken by a tiny crater.

Chevallier
44.9°N, 51.2°E

A crater lying east of Atlas A, and with a diameter of 52 km. The floor of the crater is almost fully flooded by lava. On Chevallier's floor, which is completely flat and structure-less, lies Chevallier B (13 km), a small, fresh and deep crater. The northern crater wall is breached by Chevallier H (16 km), which is also filled to the rim with lava.

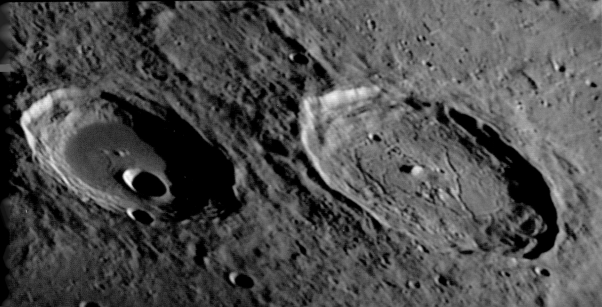

Montes Taurus 28.4°N, 41.1°E ①

The Taurus Mountains is a mountainous region east of Mare Serenitatis, extending for about 200 km. The highest peaks reach 3 km in height. The region has no clear boundaries and lies between Sinus Amoris (the 'Bay of Love') and Lacus Somniorum ('Lake of Dreams'). The conspicuous crater Römer lies within the mountains. West of Römer, and running towards the south for 110 km, are the linear rilles Rimae Römer. North of Römer there is the rille system Rimae Römer G. Both systems of rilles are clearly visible under low illumination and are observable in small telescopes.

Lacus Somniorum

Posidonius

Rimae G. Bond

Rimae Römer

le Monnier

Mare Serenitatis

Rimae Littrow

Rimae Littrow

Catena Littrow

Clerke

Mons Argaeus

Maraldi

Maraldi E

Dawes

Vitruvius

Maraldi D

Mare Serenitatis

Rima Jansen

Jansen

Vitruvius G

Lucian

Mare Tranquillitatis

Rupes Cauchy

Römer 25.4°N, 36.4°E ②

A conspicuous crater in the southern Taurus Mountains with a diameter of 40 km, a sharp rim, terraced inner walls and a central peak running north-south. There is a small crater on the northern crater wall.

Littrow 21.5°N, 31.4°E ③

A lava-flooded crater, 30 km in diameter. The crater rim is flat, eroded and, towards the south, practically non-existent. Nearby, on the west, lying between Littrow and the small crater Clerke (7 km, 21.7°N, 29.8°E) are the arcuate, Rimae Littrow rilles, with an overall length of 120 km. North of Clerke is Catena Littrow, a short chain of small craters.

Gardner 17.7°N, 33.8°E

A smaller crater, 18 km in diameter, with a smooth floor. It lies on a domed plateau, which may probably be classed as a megadome. The plateau's shape is like that of a peninsula, and is crossed by narrow rilles, chains of hills and rocky ridges.

Lying east of the plateau are the craters Maraldi, Maraldi E and Maraldi D in a north-south chain. Halfway between the craters Lucian and Vitruvius G is an un-named lunar dome. The narrow, winding rille Rima Jansen ends at the ghost crater Jansen, south of Vitruvius (30 km, 17.6°N, 31.3°E). Rima Jansen is a sinuous rille. The whole of the region around Gardner and Jansen (which should not be confused with Janssen in the southern highlands), is an interesting region for observation with all sizes of telescope.

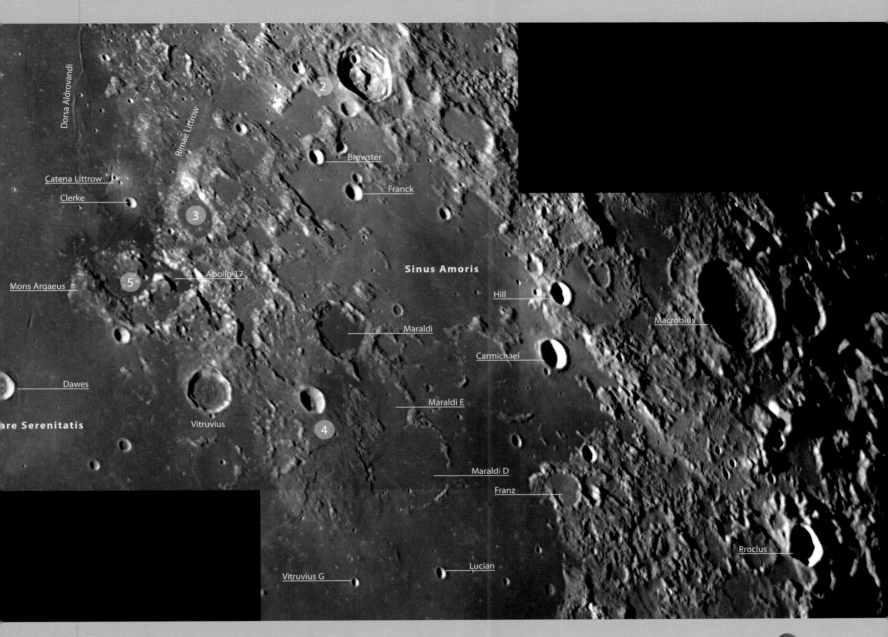

Ching-Te 20.0°N, 30.0°E

Ching-Te is a Chinese male name and is the designation for a small crater, about 4 km in diameter, near Mons Argaeus, a 50 km long mountain massif on the border of Maria Tranquillitatis and Serenitatis. The mountain is a difficult object to observe with small telescopes, especially when the angle of illumination is unfavourable. (The closer it is to the terminator, the greater the chance of being able to observe it.) Mountain massifs lie to the north, south and east of the crater.

On 11 December 1972, Apollo 17 landed about 18 km east of the crater. That site lies at the edge of the outer wall of the Serenitatis basin, in a valley between two high mountain peaks, the so-called North and South Massifs. Under good seeing conditions a small, triangular feature of bright material is visible on the dark lava surface, north of the South Massif. From terrestrial observations this feature was thought to be a landslide. Radiometric dating of Apollo 17 soil samples from this area, however, revealed a surprise. Although the rocks of the two mountainous massifs are about 3.87×10^9 years old, the samples of the bright material were only about 100 million years old. It is highly probable that it consists of ejecta from the crater Tycho, nearly 2000 km distant, that struck the massif and then slid down the slope. Another surprise from the Apollo 17 mission came from soil samples from the DMD area, which, after laboratory analysis, could unequivocally be attributed to pyroclastic volcanism.

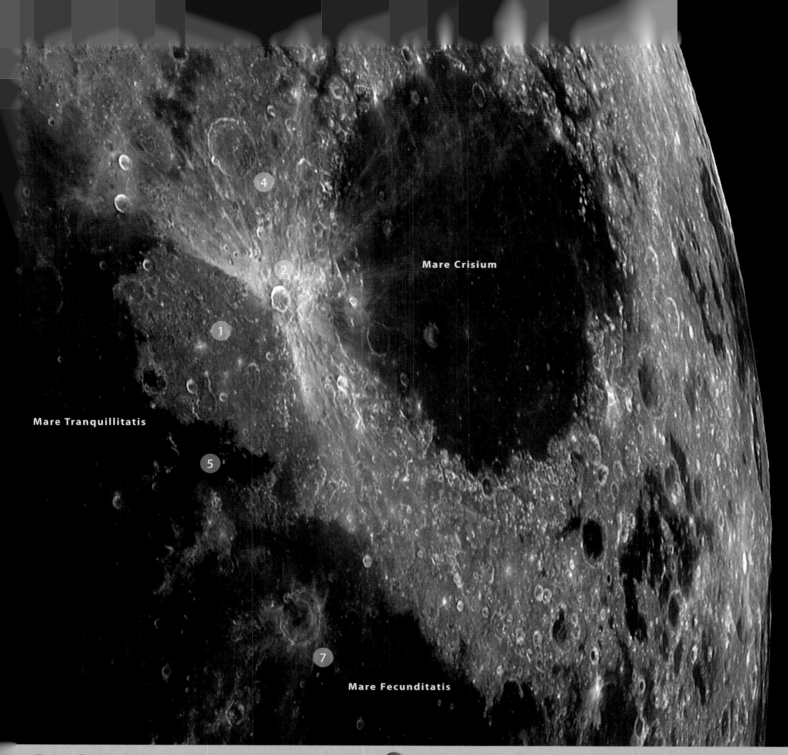

Palus Somni 14.1°N, 45.0°E ①

Palus Somni, 'Marsh of Dreams', is a diamond-shaped area, lying immediately west of Proclus. It is completely free from ejecta from the Proclus impact, and the ray system separates Palus Somni from Mare Crisium. Palus Somni extends c. 150 km in an east-west direction and appears to be an intermediate geological stage between a highland and a mare floor. The northern portion is rough and fissured, whereas the southern portion is relatively flat, smooth and level. Palus Somni shows conspicuous nuances of colour, which were formerly the subject of numerous geological studies.

Proclus 16.1°N, 46.8°E ②

Proclus is a very conspicuous, obviously five-sided crater 28 km in diameter. It lies in the western highlands, adjoining Mare Crisium. The crater walls are sharply defined and seem to be little eroded. The floor of the crater is rough and fissured, and the crater walls reach a height of 2.4 km above the crater floor. Proclus is undoubtedly one of the youngest craters. The asymmetric shape of the impact ejecta is visible when the Moon is 5 to 6 days old, and the illumination around Full Moon causes the ray system to be one of the brightest objects on the lunar surface. The longest ray runs northeast, right across Mare Crisium and is c. 600 km long (that is roughly the distance between Boston and Washington).

The ray system of Proclus changes its appearance as the age of the Moon increases. The asymmetric shape of the ray system enables us to conclude that the impactor had an almost grazing impact with an approach from the southwest. Impacts that occur at such low angles characteristically have very restricted ray systems, with asymmetric results. The lower the impact angle, the narrower the regions in which the ejecta are distributed.

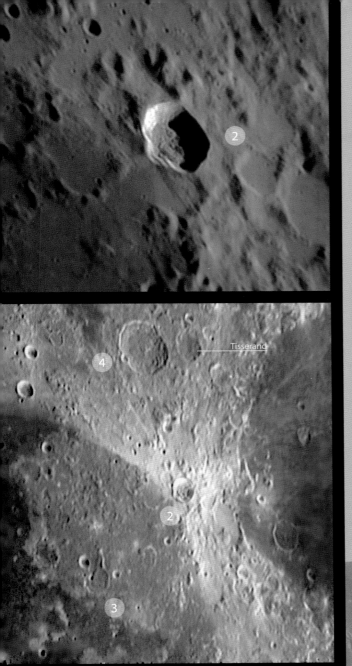

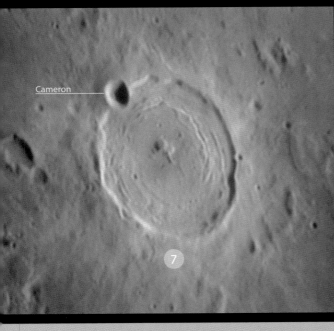

Proclus G 12.7°N, 42.7°E

Proclus G is a crater 33-km across, lying in the centre of Palus Somni. The northern crater wall opens into a conspicuous, narrow V-shaped valley.

Macrobius 21.3°N, 46.0°E

A medium-sized impact crater with a diameter of 64 km, terraced inner walls and a central peak. The western crater wall has been broken by a small crater. The nearby crater Tisserand (21.4°N, 48.2°E), with a diameter of 36 km appears like a smaller version of Macrobius. Northwest of Tisserand lies an irregular small mare surface, Lacus Bonitatis ('Lake of Goodness'). Lacus Bonitatis (23.0°N, 44.0°E) has a diameter of about 130 km.

Sinus Concordiae 10.8°N, 43.2°E

The 'Bay of Harmony' is a nondescript area – rather like a small mare – that has an east-west extent of c. 140 km. It lies at the southern end of Palus Somni.

da Vinci 9.1°N, 45.0°E

A heavily destroyed and eroded crater, with a diameter of 38 km and a rough, irregularly surfaced crater floor.

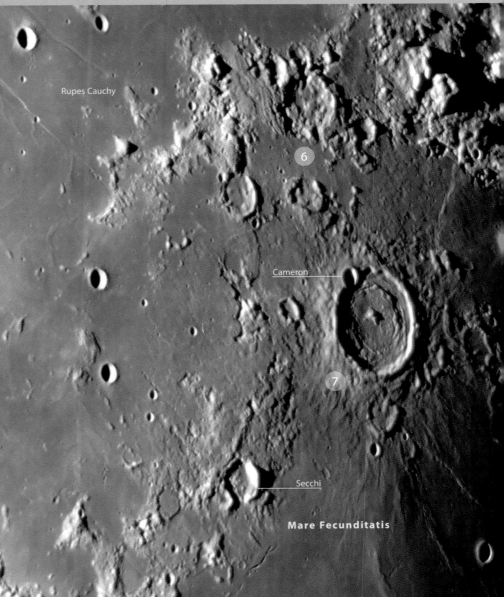

Taruntius 5.6°N, 46.5°E

A 56-km crater with a central peak and a ray system, lying between Mare Fecunditatis and Mare Tranquilitatis. Parts of the crater floor are very dark. There are landslides on the eastern, southern and western parts of the crater's floor, that almost make it appear that the crater has a double wall.

Mare Fecunditatis

Mare Crisiu

Mare Tranquillitatis

Taruntius

Anville

4

2

Dorsa Geikie

Lubbock

Lindbergh

Rimae Gutenberg

1

Ibn Battuta

Capella

Dorsa Mawson

6

7

Bellot

Montes Pyrenaeus

Crozier

Mare Nectaris

Colombo

Mare Fecunditatis **1**
5.5°S, 53.0°E

Mare Fecunditatis, 'Sea of Fertility', is an irregularly shaped lava expanse, about 500 × 600 km with an area of 300 000 km² that is thus somewhat smaller than the Caspian Sea on Earth. Because of limb-shortening through perspective, from Earth it appears oval. Like all the maria, Fecunditatis is also the lava-flooded central region of a larger impact basin and the centre of a mascon. Under high solar illumination it may be seen that the lava surface is covered with bright rays of ejecta material, which cross one another and are of different origins.

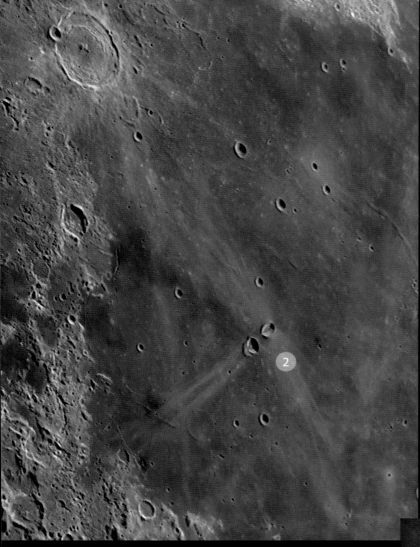

Messier 1.9°S, 47.6°E ②
Messier A 2.0°S, 47.0°E

Messier and Messier A together form one of the most striking crater pairs on the nearside of the Moon. Messier is a markedly oval crater that measures about 9 × 11 km. Its oval shape is the true one, and is not caused by perspective through limb foreshortening.

Messier A is a double crater with overall measurements of about 11 × 13 km. The two craters overlap. Messier A is the origin of two bright rays of ejecta, each of which is c. 100 km long, and which look like the tail of a comet. The two craters were possibly created simultaneously through the grazing impact of one or two bodies, although there was possibly just one impactor, which broke into two parts through gravitational effects, shortly before impact. It is uncertain whether Messier and Messier A were created simultaneously. Lying east and south of Messier and prominent under low solar illumination are the mare wrinkle ridges Dorsa Geikie (240 km, 3°S, 53°E) and Dorsa Mawson (180 km, 7°S, 53°E).

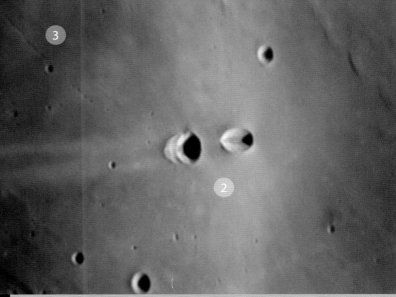

Rima Messier 1.0°S, 45.0°E ③

A small linear rille about 100 km long, lying northwest of Messier. The rille's maximum width is 1 km, so requires a large aperture to be observed.

8b Mare Fecunditatis

Secchi 2.4°N, 43.5°E

A smaller crater with a diameter of 22 km, with a wall that is broken in the north and south. The western crater wall adjoins the small and not very pronounced range of mountains, known as Montes Secchi (3.0°N, 43.0°E, length c. 50 km).

Rimae Secchi 1.0°N, 44.0°E

Rima Secchi is a difficult-to-observe rille system, c. 35 km long, lying south of the crater Secchi. At the northeastern end, the rille splits into the form of a 'Y'. An aperture of at least 20 cm is needed for it to be observed.

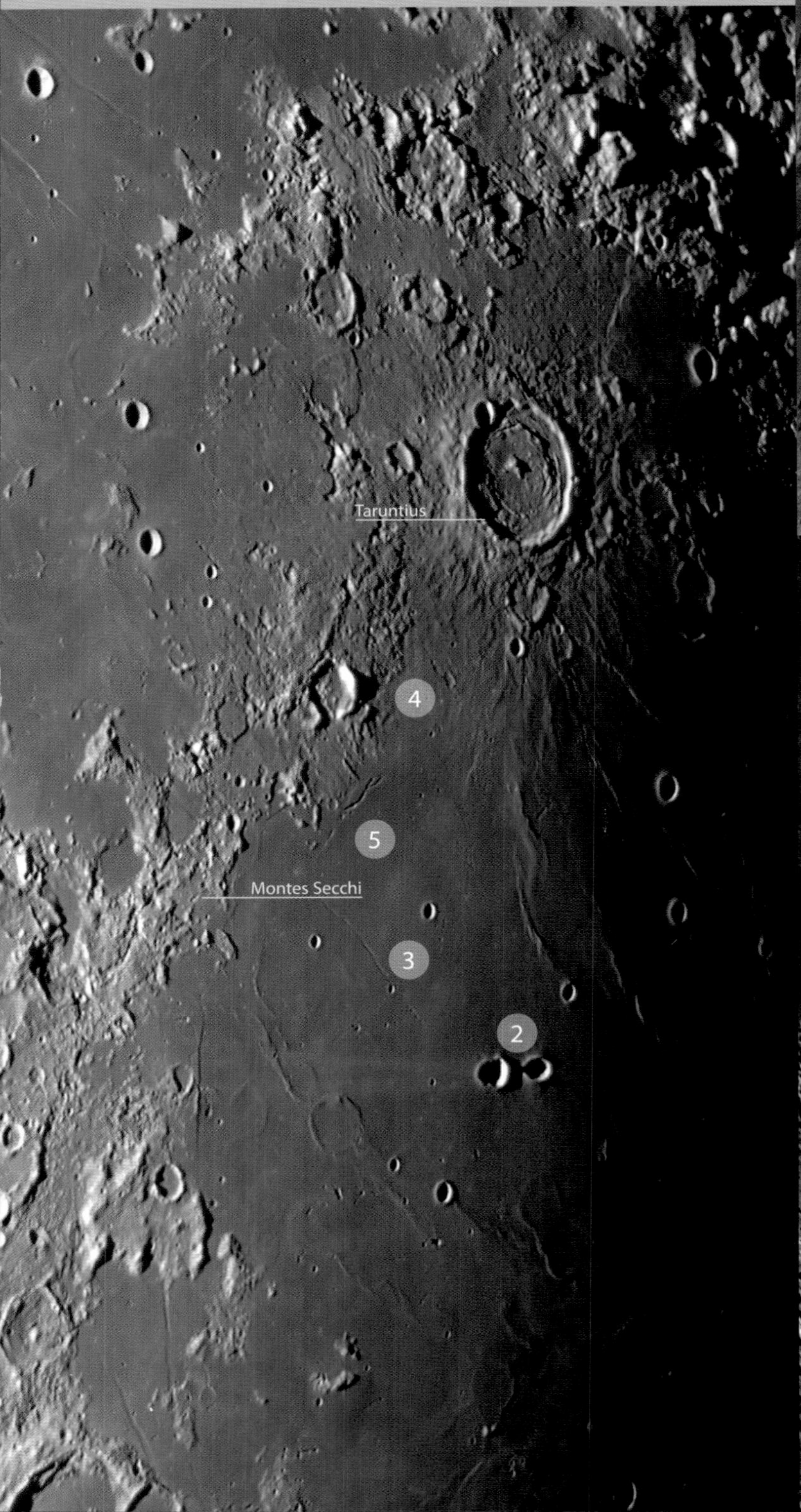

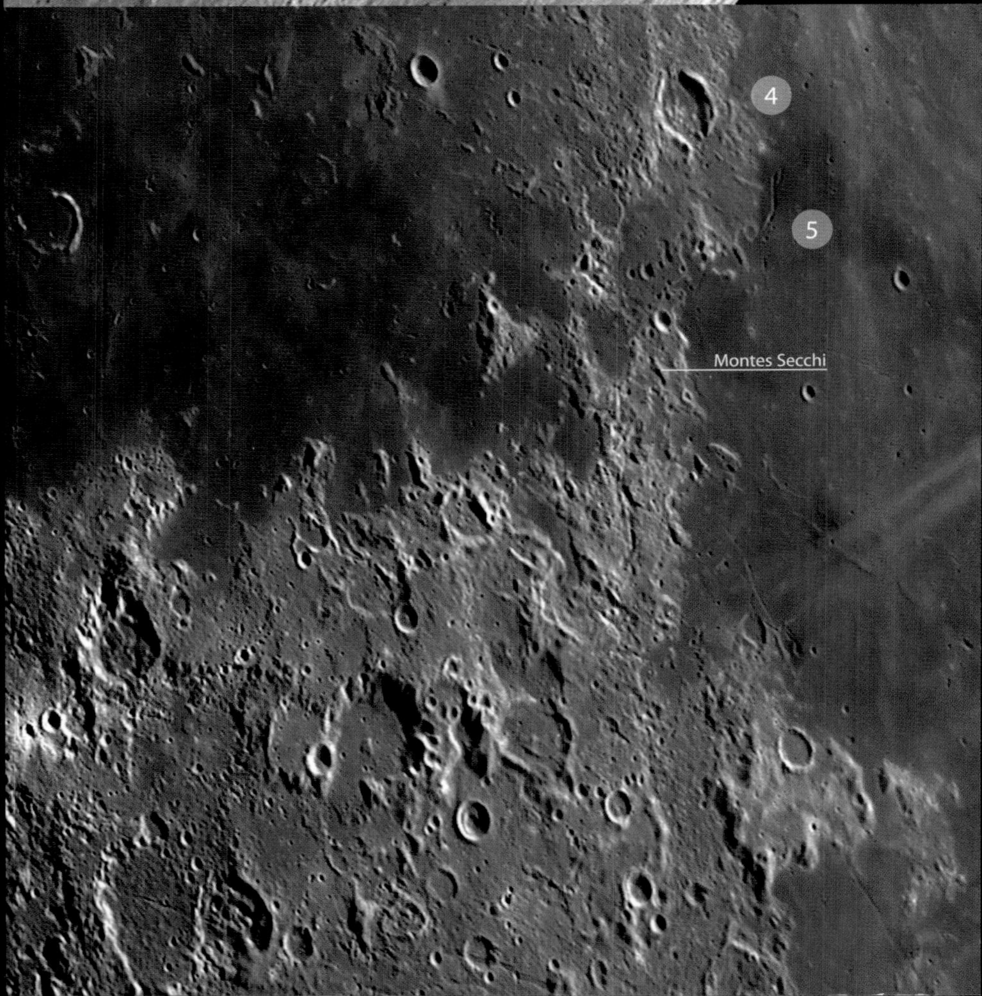

Langrenus

Taruntius

Montes Secchi

Montes Secchi

Gutenberg 8.6°S, 41.2°E ⑥

A lava-flooded crater, 74 km in diameter. The eastern wall is broken by the similarly lava-flooded crater Gutenberg E (28 km). To its south lies Gutenberg C (45 km), which has completely destroyed the southern wall of Gutenberg itself. On the southwestern wall lies Gutenberg A (15 km). Within the crater a series of hills and the southern portion of Rimae Gutenberg are visible.

The rille system has a total length of c. 330 km. To the east lies the Rimae Goclenius rille system, which has an overall length of 240 km. Both systems consist of linear, largely parallel rilles, which are wide enough to be seen even with smaller telescopes. Each of the individual rilles consists of two parallel fault zones, with a sunken floor lying between them. They probably arose when the massive lava surface of Mare Fecunditatis sank under its own weight when it cooled.

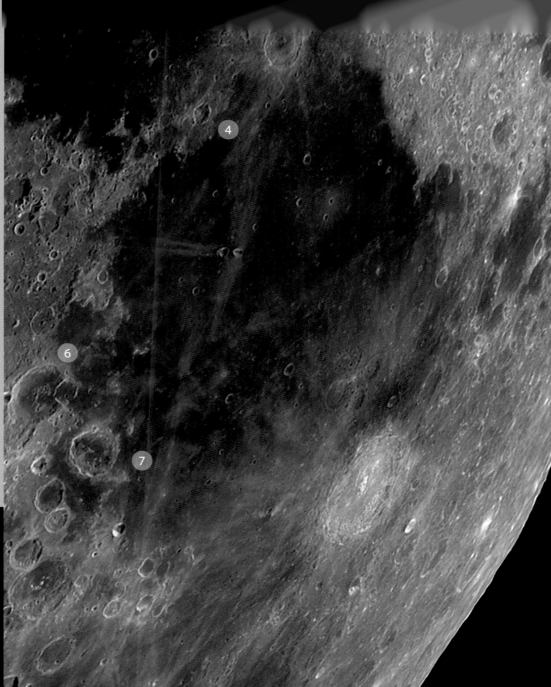

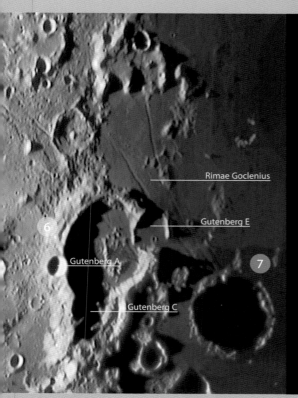

Rimae Goclenius

Gutenberg E

Gutenberg A

Gutenberg C

Goclenius 10.0°S, 45.0°E ⑦

Goclenius is an extremely irregularly shaped crater, measuring 54 × 72 km. An extension of Rimae Goclenius (8.0°S, 43°E) crosses the crater floor. The remainder of a respectable rille system extends over a distance of 240 km in a northwesterly direction. The rille system thus follows the western edge of Mare Fecunditatis. When illuminated after First Quarter, the crater wall appears bright.

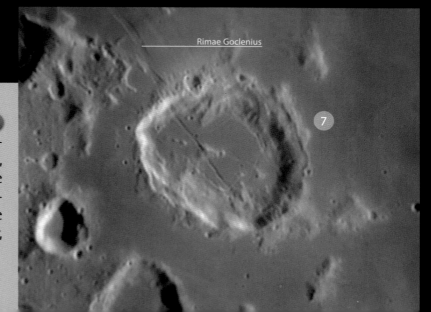

Rimae Goclenius

Mare Fecunditatis

Lohse

Lamé

Holden

Petavius B

Vallis Snellius

Langrenus 8.9°S, 61.1°E

Langrenus is a very large and striking complex crater in the southwestern quadrant of the Moon. It has a diameter of 127 km, with 2.7 km-high, terraced inner crater walls, central peaks and a hilly crater floor. Under low solar illumination, the floor of the crater appears to have yellowish-brownish tints, in comparison with the neutral grey of the crater's surroundings. Langrenus has two conspicuous central peaks, which on old Moon maps were designated Langrenus alpha (the southern) and Langrenus beta. The southern crater wall is shaped rather like a triangular cape, and below it lie the craters Langrenus O and E, and Lohse (42 km, 13.7°S, 60.2°E).

Petavius 25.1°S, 60.4°E

Petavius is a giant complex crater with a diameter of 190 km. The inner crater walls are terraced and tower almost 3.3 km above the crater floor. Fairly central lies a large central, mountainous massif (south of which is the craterlet Petavius A).

On the floor of Petavius, a fracture zone (the main rille of Rimae Petavius) runs through the central massif. An additional segment begins on the western side of the central massif and runs farther west to the inner crater wall, and from there – less distinct – in both north and south directions, thus following the crater wall. These clefts and faults, together with the dark patches on the floor of the crater, indicate post-volcanic activity, long after the actual impact event.

North of Petavius lies the c. 30 km crater Petavius B (20.0°S, 57.0°E). It displays a bright, asymmetrical ray system, which stretches over a large area of Mare Fecunditatis. East of Petavius B, the system seems to fan out.

Vendelinus 16.4°S, 61.6°E

A large complex crater, 140 km in diameter. The eastern wall has been interrupted by the crater Lamé (14.7°S, 64.5°E), 84 km in diameter, and other small craters. The 47-km crater Holden (19.1°S, 62.5°E) lies on the southern wall. The smaller craters Vendelinus H, E, L, Z and Y (10 km to 17 km across) are visible on the crater's floor.

Snellius 29.3°S, 55.7°E

Snellius is a large crater, 83 km in diameter, lying southwest of Petavius. The crater is located in the centre of Vallis Snellius, which runs across Snellius from northwest to southeast.

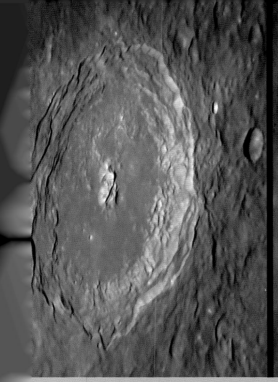

Rimae Petavius

Petavius A

Furnerius 36.0°S, 60.6°E (5)

Furnerius, with its diameter of 135 km and a strongly eroded northern crater wall is among the largest complex craters in the southeastern highlands. As well as a large, dark lava surface, the crater floor contains the large, eccentrically placed crater Furnerius B (22 km) and, to its east a rille, c. 50 km long, that stretches as far as the northern inner crater wall. Outside the northern wall, halfway to the crater Adams (64 km, 31.9°S, 68.0°E), is the beginning of Vallis Snellius (31.0°S, 59.0°E), which, with its overall length of c. 590 km, is one of the longest lunar valleys on the nearside. It runs in the direction of the centre of the Mare Nectaris basin, and is probably related to the latter's formation. The valley consists of a multitude of overlapping craters of various diameters. Like Vallis Rheita it probably arose through secondary impacts, initiated by the gigantic Nectaris impact.

Stevinus 32.5°S, 54.2°E (6)

Stevinus is a conspicuous crater, 75 km in diameter, northwest of Furnerius. It has terraced inner walls and a central peak. Directly west lies the small crater Stevinus A (c. 6 km, 32.0°S, 52.1°E). Under high solar illumination the crater exhibits bright rays of ejected material. The craters Snellius and Furnerius A are also the centres of bright ray systems.

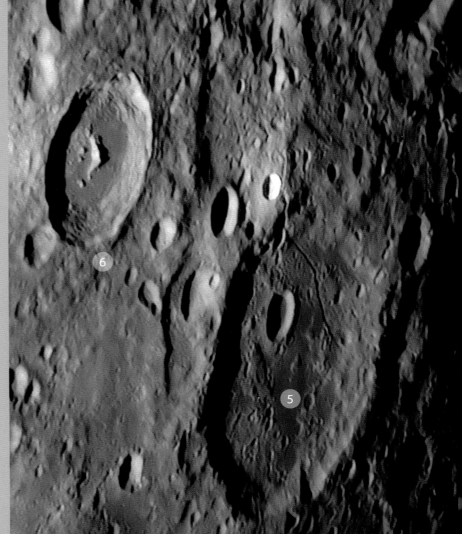

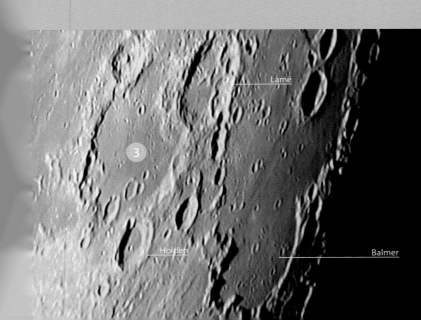

Lamé

Holden

Balmer

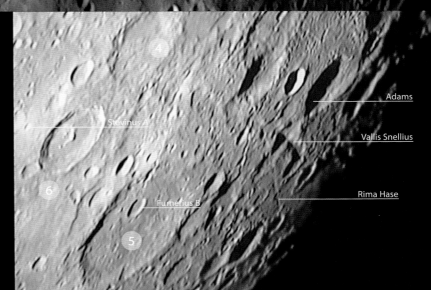

Adams

Stevinus A

Vallis Snellius

Rima Hase

Furnerius B

Mare Australe
50.0°S, 93.0°E

Mare Australe, the 'Southern Sea', lies on the southeastern limb of the Moon. Because of limb foreshortening, details are difficult to observe and a large portion lies on the lunar farside. It is a large impact basin. Lunar Orbiter images show an almost perfectly circular structure with a diameter of over 600 km and an area of roughly 320 000 km^2.

Images from lunar probes show, in addition, over 200 craters, with a wide range of sizes. Mare Australe is probably one of the oldest basins and was subsequently only partially flooded with mantle lavas. So there are no traces of the basin's rims, and neither can ejecta from craters that were created later be detected. As a result, the diameter is very difficult to determine, with figures, depending on their source, varying between 500 and 900 kilometres. The Australe basin was probably created before the Late Heavy Bombardment, and the walls of the basin were completely destroyed by subsequent impacts. That would also explain the atypical number of large craters for a mare that occurs in this region.

Hanno 56.3°S, 71.0°E
Pontécoulant 58.7°S, 66.0°E
Oken 43.7°S, 75.0°E

Under good libration conditions, the craters Hanno (55 km), Pontécoulant (90 km) and Oken (70 km) may be readily located. All three stand out because of their dark crater floors (darker than the mare lava). Pontécoulant has a few small craters on its floor, but the floors of Hanno and Oken are smooth and flat.

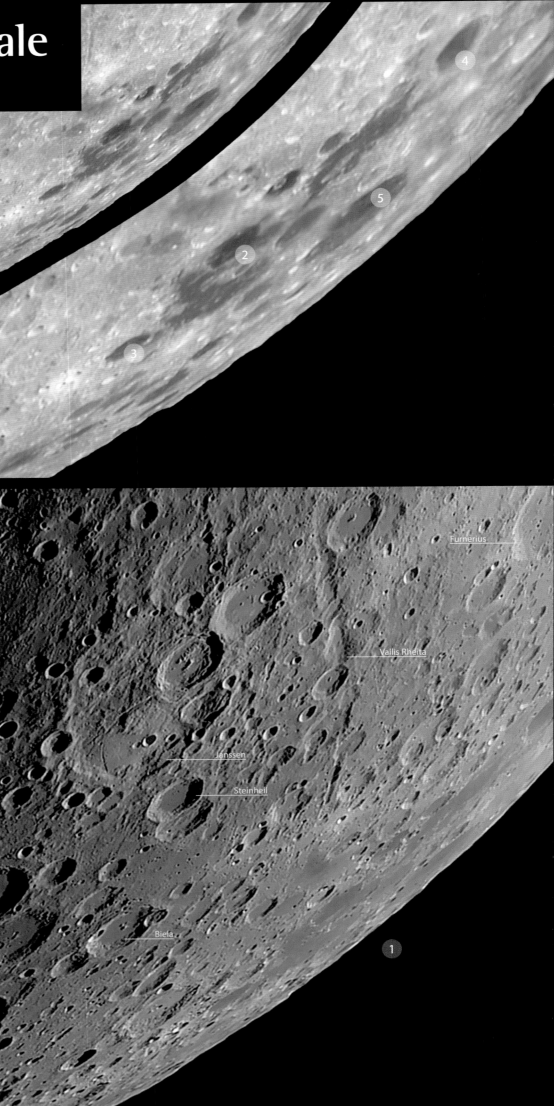

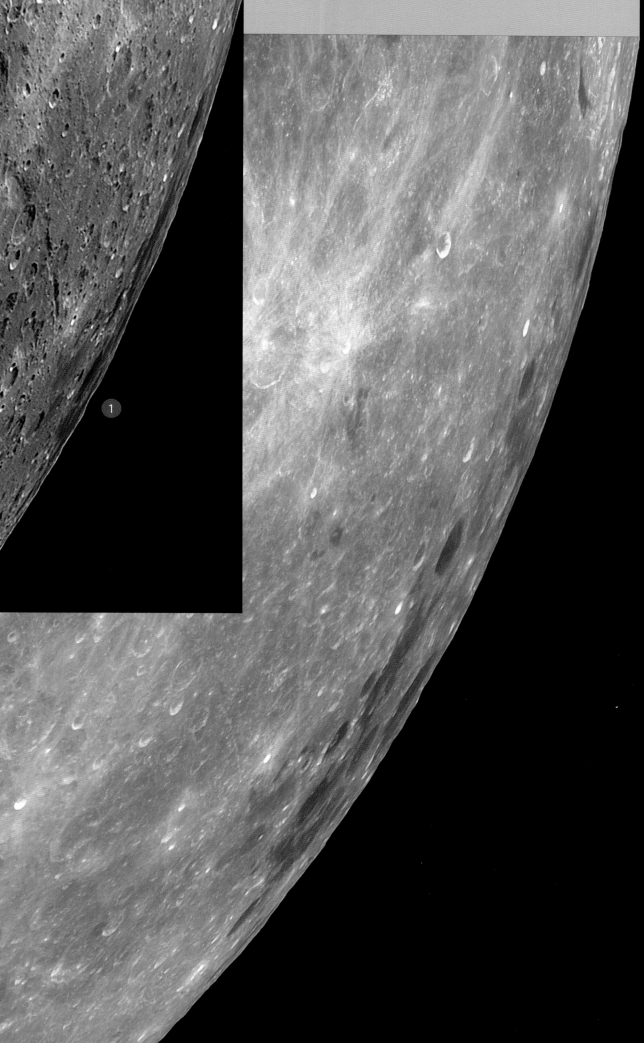

Lyot 50.2°S, 84.1°E

Lyot, with a diameter of 140 km, is a large complex crater and the largest impact structure in Mare Australe. Observing it is a challenge because of its location at the extreme limb and is successful only under the most favourable libration.

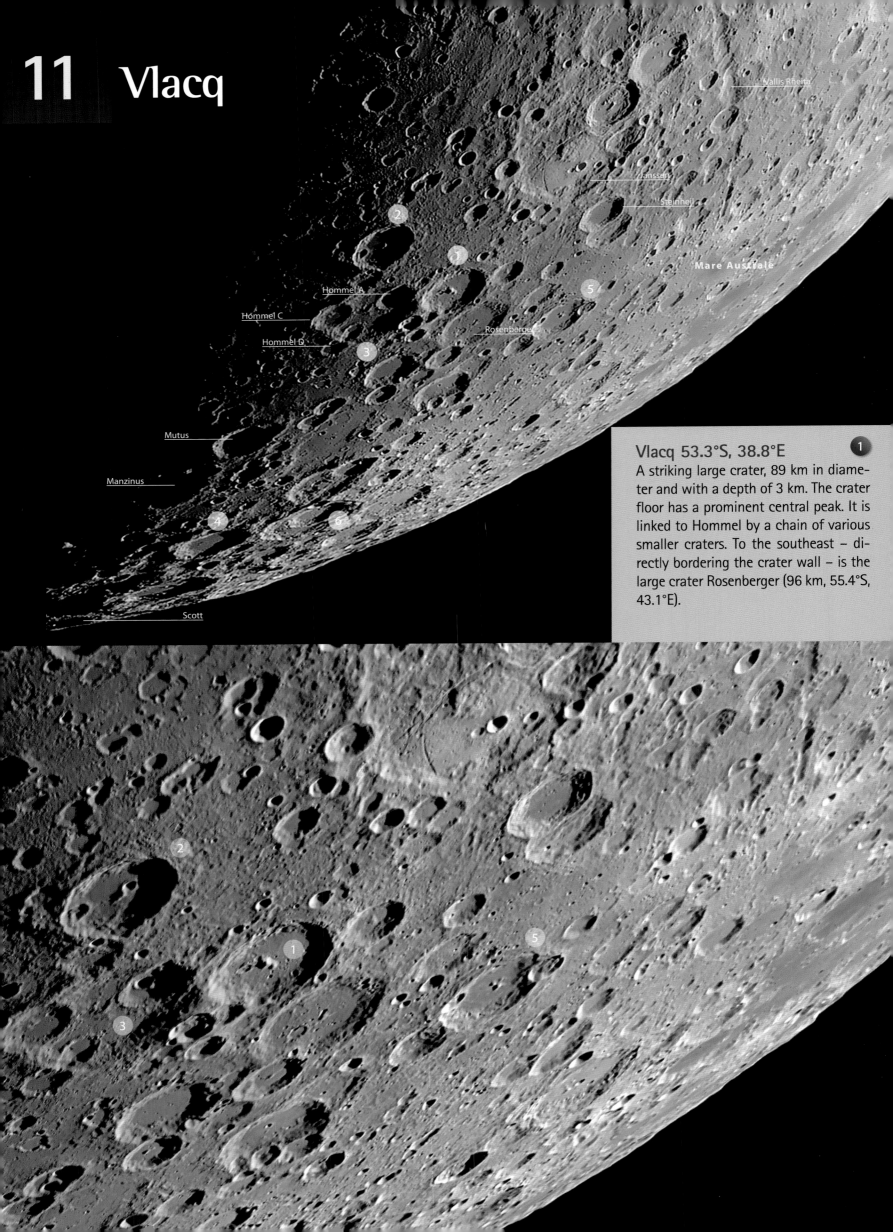

Vallis Rheita

Janssen

Steinheil

Mare Australe

Hommel A

Hommel C

Hommel D

Rosenberger

Mutus

Manzinus

Scott

Vlacq 53.3°S, 38.8°E ❶
A striking large crater, 89 km in diameter and with a depth of 3 km. The crater floor has a prominent central peak. It is linked to Hommel by a chain of various smaller craters. To the southeast – directly bordering the crater wall – is the large crater Rosenberger (96 km, 55.4°S, 43.1°E).

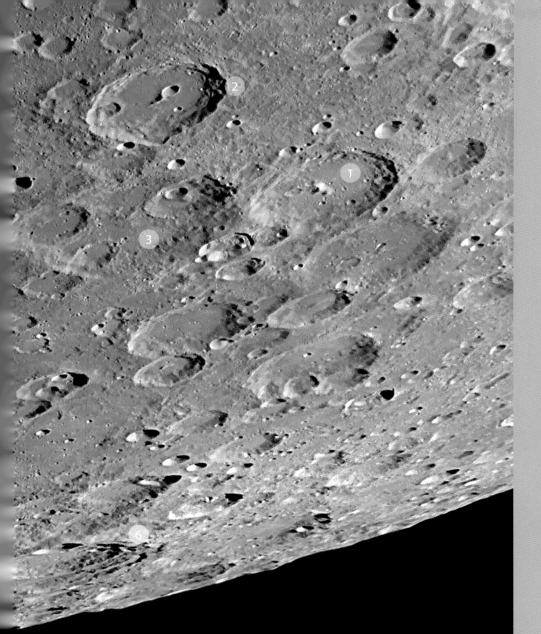

Pitiscus 50.4°S, 30.9°E ②

Like Vlacq, also a prominent crater, 82 km in diameter with a flat, lava-flooded crater floor. On the floor lies Pitiscus A (10 km) and superimposed on the western rim is Pitiscus E (13 km).

Hommel 54.7°S, 33.8°E ③

An ancient, fissured crater, heavily eroded by later impacts, that is 126 km in diameter. It was created in the Pre-Nectarian epoch on the lunar timescale. The crater floor lies 2.8 km below the crater's rim and on the floor and on the western wall there are three large craters, Hommel A, D and C (50 km, 28 km and 53 km, respectively). Hommel is bordered directly to the east by Vlacq and to the north by the large crater Pitiscus.

Hommel, Vlacq and Pitiscus form an interesting group, but because of their limb location are not easy to observe. Observations under favourable libration angles are recommended.

Boguslawsky 72.9°S, 43.2°E ④

A large crater, with a diameter of 97 km and a depth of 3.4 km, which lies west of Boussingault. The crater walls seem to be eroded, broken down, and not sharply defined, similar to many large craters in the southern highlands. Against the inner, eastern crater rim lies Boguslawsky D (24 km). Both craters lie well towards the limb, and require favourable libration angles to be observed.

Biela 54.9°S, 51.3°E ⑤

A large crater with a diameter of 76 km, a smooth crater floor, and with prominent central mountains.

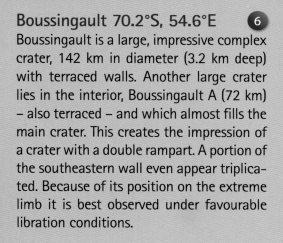

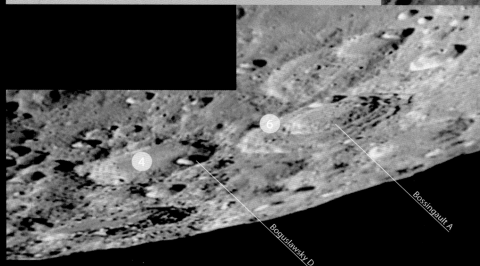

Boussingault 70.2°S, 54.6°E ⑥

Boussingault is a large, impressive complex crater, 142 km in diameter (3.2 km deep) with terraced walls. Another large crater lies in the interior, Boussingault A (72 km) – also terraced – and which almost fills the main crater. This creates the impression of a crater with a double rampart. A portion of the southeastern wall even appear triplicated. Because of its position on the extreme limb it is best observed under favourable libration conditions.

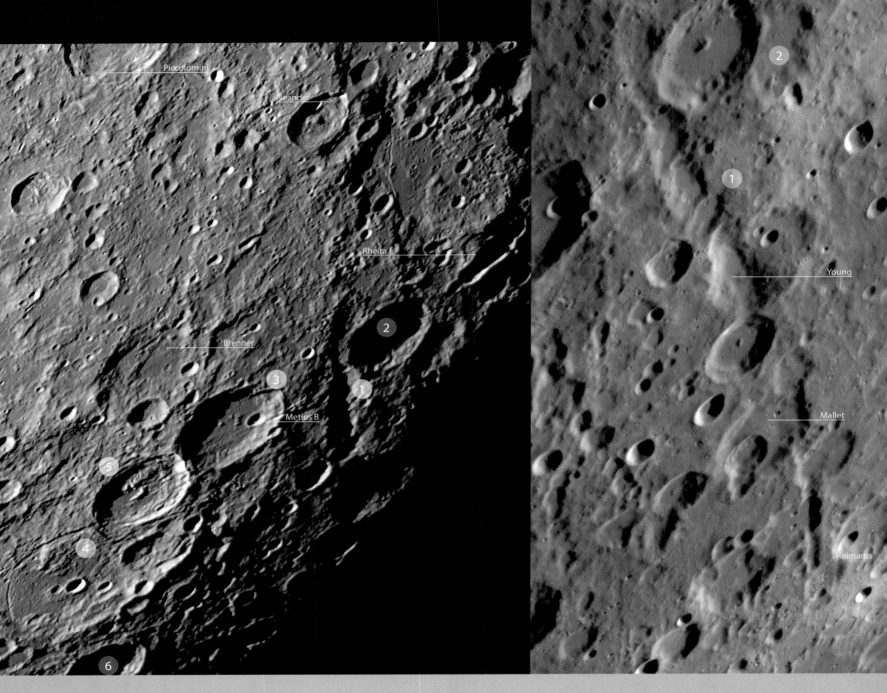

Piccolomini
Neander
Rheita E
Brenner
Metius B
Young
Mallet
Reimarus

Vallis Rheita 42.5°S, 51.5°E ①

Vallis Rheita is a linear valley, approximately 450 km long (similar to Vallis Snellius). At its widest points it reaches a width of about 30 km. It consists of a string of numerous, partially overlapping craters. It is oriented directly towards the centre of the Nectaris Basin and it is suggested that it may have arisen from secondary events following the Nectaris impact.

The crater chain begins northwest of the crater Rheita, crosses the crater Young (71 km, 41.5°S, 50.9°E), then the crater Mallet (58 km, 45.5°S, 54.2°E) and ends southeast of the crater Reimarus (48 km, 47.7°S, 60.3°E).

Rheita 37.1°S, 47.2°E ②

Rheita is a prominent crater, 70 km in diameter and with an eccentrically placed central peak, and lies northeast of Metius. Vallis Rheita runs past the crater's southern wall. Slightly farther north lies the crater Rheita E, which, with its measu-

rements of 66 × 30 km, exhibits a very elongated shape. It consists of at least three overlapping individual craters, which probably (like Schiller) gained its current form through a later impact.

If the position of Janssen, Fabricius, Metius, Rheita and Rheita E is examined at a large scale, then it appears that they all lie on a nearly straight line. It is conceivable that their formation, nearly simultaneously, may be traced back to a very large asteroid that broke apart shortly before the impact.

Metius 40.3°S, 43.3°E ③

The crater lies immediately north of Fabricius. It has low central peaks and the crater Metius B (14 km) lies on the northeastern portion of the floor, which is otherwise relatively flat and level. Metius, with a diameter of 85 km is slightly larger than Fabricius. The outer walls of Janssen, Fabricius and Metius are in direct contact with one another.

Janssen 45.4°S, 40.3°E

Janssen – not to be confused with Jansen in Mare Tranquillitatis – is a giant complex crater that is c. 200 km in diameter, and has a wide rille system and central mountain massif. It has been considerably damaged and altered by secondary impacts originating with the Nectaris impact. Janssen must, therefore, be significantly older than the Nectaris Basin. The system of arcuate rilles in the southern portion is known as Rimae Janssen. With a length of about 120 km and a maximum width of 3 km, they are visible even in small telescopes. They appear like tension cracks, because they follow the track of the crater wall, but their actual origin is uncertain and unique in a large complex crater. The northern wall is broken by the large crater Brenner (97 km, 39.0°S, 39.3°E). Brenner is greatly eroded, the craters Fabricius and Metius are much more distinct. Janssen should perhaps be included in the class of FFC craters.

Fabricius 42.9°S, 42.0°E

Fabricius lies within the northern part of Janssen. The crater has a diameter of 78 km and an elongated central peak, rich in structure. Observations with smaller telescopes suggest the image of a crater with a double wall, as if two craters were lying concentric with one another. Larger telescopes show, however, that this is an optical illusion, because the supposed, second, inner crater wall appears on the west, north, and east as a continuous horseshoe-shaped structure, consisting of mountain ridges and landslides.

Steinheil 48.6°S, 46.5°E
Watt 49.5°S, 48.6°E

These are a striking pair of overlapping craters, with diameters of 67 km, but of very different morphology. The floor of Steinheil is flat, level and flooded by lava. The eastern, inner crater wall is topped by a small, but clearly visible crater. The floor of Watt appears rough and wavy. A mountain ridge divides the floor of the crater in a north-south direction.

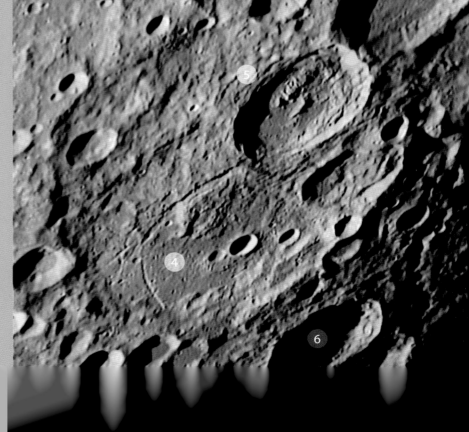

Mare Nectaris
15.2°S, 35.5°E

Mare Nectaris (Sea of Nectar) is a relatively small and nearly circular lunar maria of about 350 km in diameter. It is the lava-flooded central portion of a large impact basin with an area of c. 100 000 km². It arose in the Nectarian epoch (about 3.9 billion years ago), when the lunar surface was created, and is one of the oldest basins on the nearside of the Moon. Nectaris is a multi-ring basin identified by the ring-like arrangement of four concentric basin ramparts. Mare Nectaris is also the centre of a gravitational anomaly (a mascon).

Smaller telescopes show very few craters on the lava surface of Mare Nectaris (the largest of them is Rosse). Larger telescopes, however, show a wealth of craterlets and pits. One interesting surface feature, northeast of the crater Rosse, is a bright ray of ejecta from the Tycho impact. Large telescope reveal that, within the ray, there are many tiny secondary craters (in a so-called crater cluster).

Daguerre
11.9°S, 33.6°E

A ghost crater, 46 km in diameter, in the northern area of Mare Nectaris. A group of crater pits lies almost exactly in the centre of the crater's floor. Its southern half is crossed by the long ray of ejecta from Mädler. Directly adjoining it to the northwest lie a second, nameless ghost crater, that is very similar to Daguerre in structure.

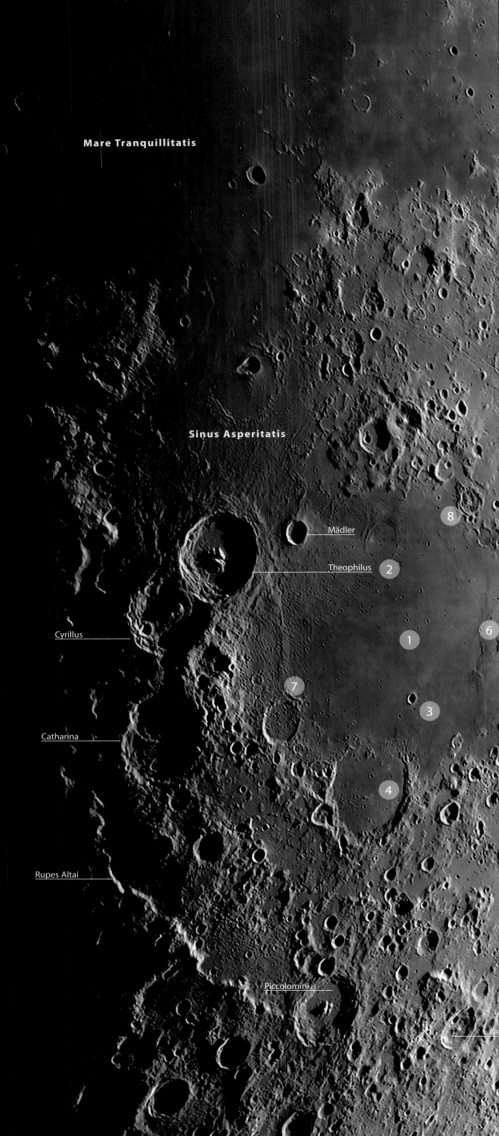

Mare Tranquillitatis

Sinus Asperitatis

Mädler

Theophilus

Cyrillus

Catharina

Rupes Altai

Piccolomini

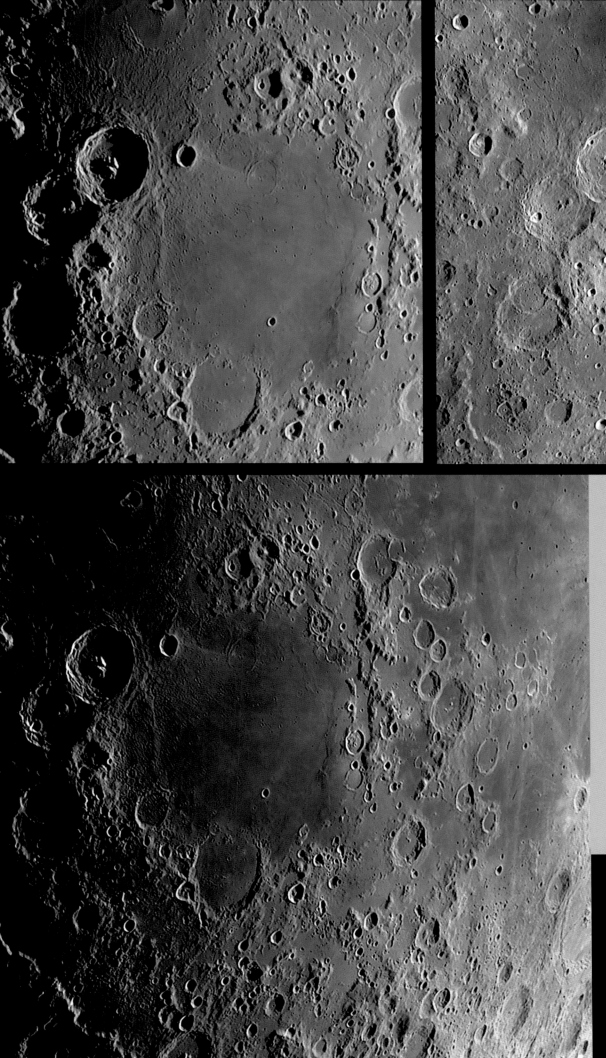

Outer rings of the Nectaris Basin

Mare wrinkle ridges	240 km
Montes Pyrenaeus	400 km
Craters Santbech/Cyrillus	620 km
Rupes Altai	860 km

The crater Piccolomini lies on the outermost basin rampart (Ring 4), and a segment of this wall is the Rupes Altai scarp. Theophilus, Cyrillus and Catharina lie between Rings 2 and 3. The system of mare wrinkle ridges marks the inner wall of the basin.

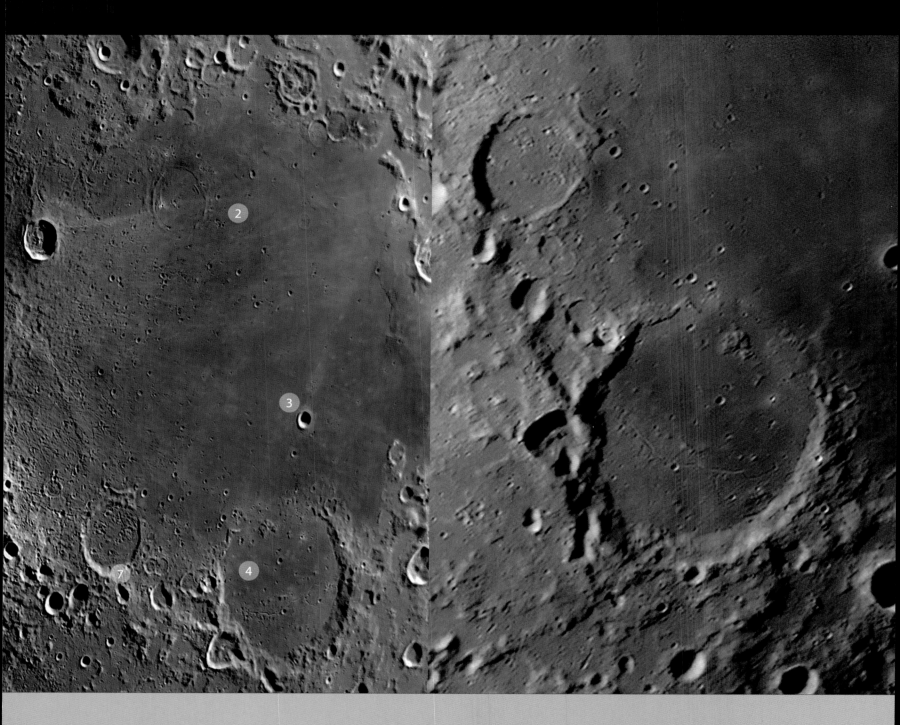

Rosse 17.9°S, 35.0°E ③

A young, circular crater, diameter 12 km, lying in the southern area of Mare Nectaris. A bright ray of ejecta from the Tycho impact runs northwards from it (from southwest to northeast). Under high solar illumination, larger instruments show a group (a crater cluster) of very bright tiny secondary craters within the ray. Rosse is the only conspicuous crater on the monotonous lava expanse of the Nectaris Basin and therefore stands out all the more.

Fracastorius 21.5°S, 33.2°E ④

Fracastorius is a highlight for any aperture of telescope. It is a large complex crater of about 120 km in diameter, that is slightly oval in the north-south direction. Like a bay, the crater wall opens in the north onto the lava surface of Mare Nectaris, in a way that is morphologically similar to the craters Letronne (Mare Humorum) and

Le Monnier (Mare Humorum). However, low remnants of the wall are still detectable in the north.

A small, nameless rille crosses the crater floor, almost exactly from west to east, and which splits into a 'Y' shape on the eastern side. It crosses the small crater Fracastorius M (4 km). The southern floor of the crater is saturated with a whole host of crater pits, but the northern floor is relatively flat and level. Well to the north is a small crater, surrounded by a bright halo of anorthositic crustal material. Larger telescopes show a small group of mountains on the eastern crater wall and, in the southern area, a few rounded, hill-like domes.

Fracastorius probably arose after the Nectaris impact. The fracture zone and the many craters on the floor were formed after the lava flows that cover the floor of Fracastorius had cooled.

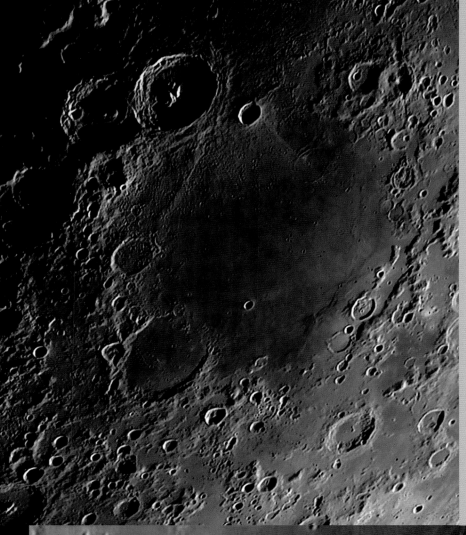

Montes Pyrenaeus 15.6°S, 41.2°E

A train of mountains about 164 km long, beginning south of the crater Gutenberg. The Pyrenees are a remnant of the inner wall of the Nectaris Basin.

Bohnenberger 16.2°S, 40.0°E ⑥

A crater of 33 km diameter with a few low, flat hills on the crater floor. It lies at the eastern end of Mare Nectaris, where, under grazing illumination, an extended mare wrinkle-ridge system is visible.

Beaumont 18.0°S, 28.8°E ⑦

A lava-flooded crater, 53 km in diameter, lying on the western edge of Mare Nectaris. The eastern wall is open to the mare floor and a low mountain ridge runs north of the crater wall.

Gaudibert 10.9°S, 37.8°E ⑧

A crater, 43 km in diameter with many sharp mountain ridges on the crater floor. A trio of small craters lies bordering the southern wall, while the rest of the surrounding crater wall partially consists of mountains and mountain chains.

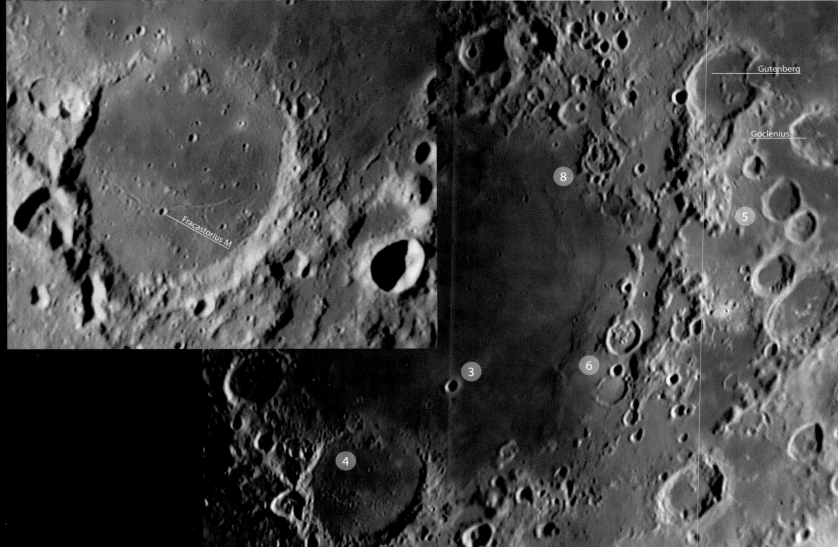

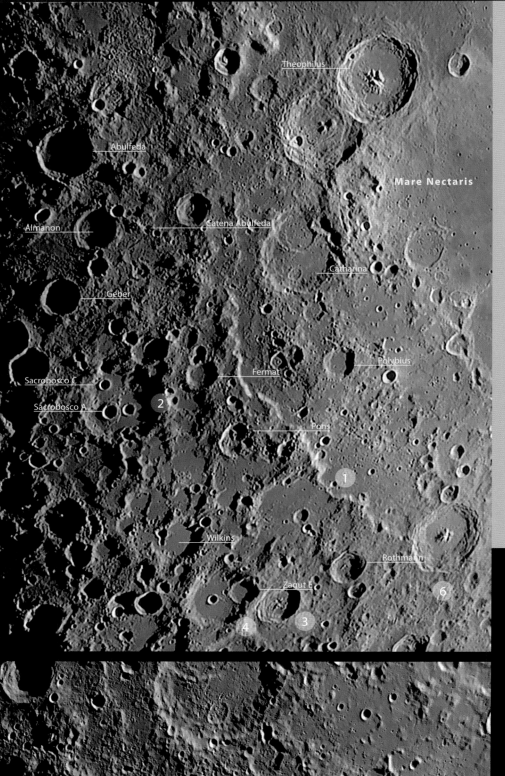

Rupes Altai 24.3°S, 22.6°E ⓵

A slightly arcuate escarpment between the craters Piccolomini and Catharina, with a length of about 440 km. The escarpment is a portion of the outer wall of the Nectaris Basin. The greatest height difference between the mountain crests and the surrounding terrain of about 3 km lies near the crater Pons (44 × 31 km, 25.3°S, 21.5°E) and Fermat (39 km, 22.6°S, 19.8°E) and west of the crater Polybius (41 km, 22.4°S, 25.6°E). The inclination of the slopes of Rupes Altai is much steeper than, for example, with Rupes Recta or Rupes Cauchy. Rupes Recta and Rupes Cauchy also have a much smaller difference in height compared with the surrounding terrain.

Sacrobosco 23.7°S, 16.7°E ⓶

An extremely strongly eroded crater with a diameter of 98 km. The crater floor is smooth and flat. Two distinctive craters, Sacrobosco A (18 km) and B (14 km) lie in the centre of the crater, and a third crater – Sacrobosco C (13 km) – right on the northern, inner edge of the crater.

Lindenau 32.3°S, 24.9°E ⓷

A distinctive crater with terraced walls and multiple central peaks. It has a diameter of about 53 km and the crater walls rise about 2.9 km above the crater's floor.

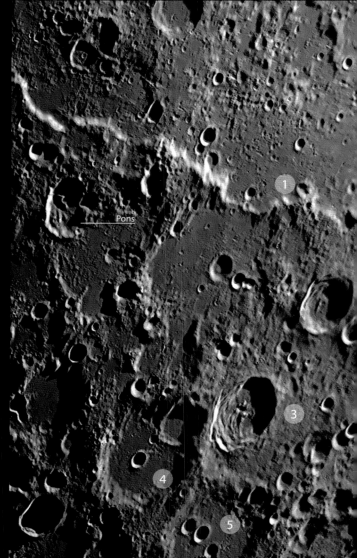

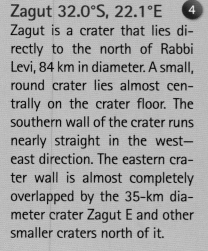

Zagut 32.0°S, 22.1°E ④

Zagut is a crater that lies directly to the north of Rabbi Levi, 84 km in diameter. A small, round crater lies almost centrally on the crater floor. The southern wall of the crater runs nearly straight in the west–east direction. The eastern crater wall is almost completely overlapped by the 35-km diameter crater Zagut E and other smaller craters north of it.

Rabbi Levi ⑤
34.7°S, 24.0°E

Rabbi Levi is likewise a prominent crater with a diameter of 81 km. On the western portion of the crater floor there are five, additional, conspicuous, but much smaller, craters. A group of smaller craters overlies the eastern crater wall.

Piccolomini ⑥
29.7°S, 32.2°E

Piccolomini is a 87-km diameter, very prominent crater with a large central peak. The inner walls are terraced and reach a height of 4.5 km above the crater's floor. It certainly arose in the late Imbrium period.

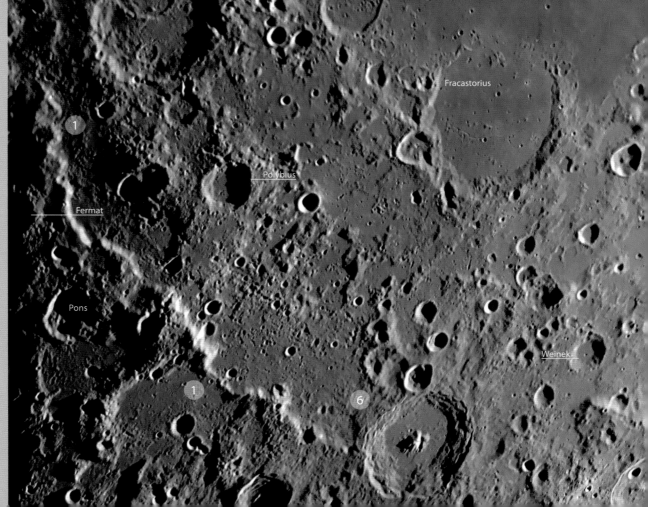

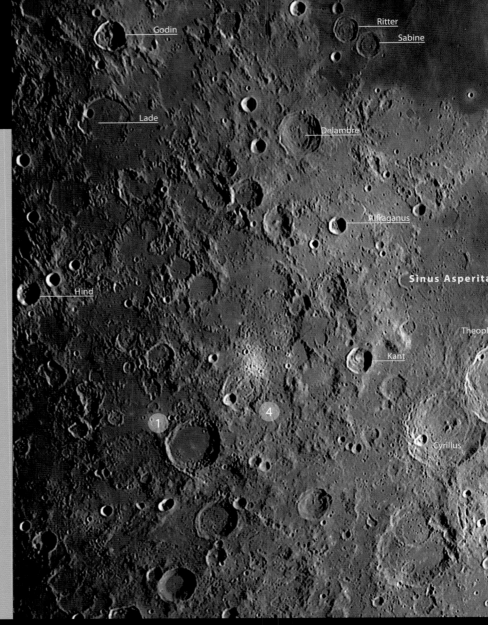

Abulfeda 13.8°S, 13.9°E ❶

Abulfeda is a very regularly formed crater with a diameter of 65 km and a depth of 3.1 km. The crater floor is relatively smooth and flat, and the southern, inner crater wall exhibits features like landslides. Fairly central there is a craterlet that is surrounded by a very bright halo of ejecta. Large telescopes show additional crater pits and hills on the crater's floor. Abulfeda is a typical crater, like many of the others that were formed during the Nectarian period (between 3.92 and 3.85 billion years ago).

Abenezra 21.0°S, 11.9°E ❷

An almost rectangular crater with a diameter of 43 km and a wall height of about 3.7 km, measured from the crater's floor. The crater Abenezra C (44 km) to the west and partly overlapped by Abenezra, forms part of a trio of craters with Azophi.

Azophi 22.1°S, 12.7°E ❸

A crater 48 km in diameter and 3.7 km deep.

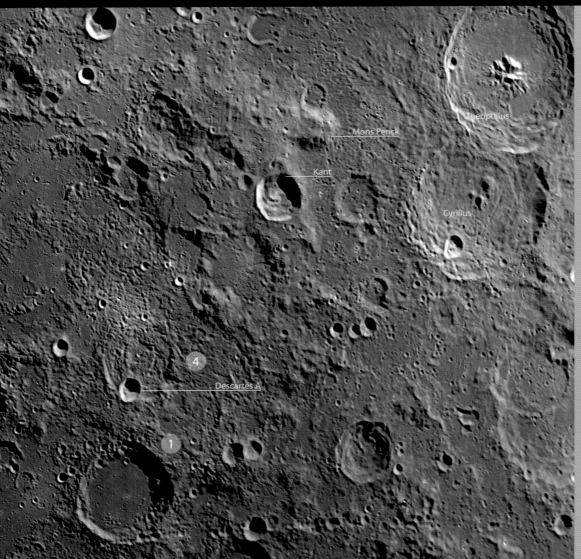

Descartes 11.5°S, 15.7°E ❹

The region around the crater Descartes is very interesting and was therefore chosen as the landing site for Apollo 16. Geologists expected soil samples to be of volcanic origin. Contrary to the lunar geologists' assumptions no rocks of any form of volcanic origin were found in the region.

Descartes is a heavily eroded crater, 48 km in diameter, and exhibits a bright area, extending out from the northern crater wall. The material consists of breccia and once molten highland material. The origin of the bright ejecta is uncertain. The crater wall of Descartes A (16 km) also appears extremely bright in sunlight. In 1999, the lunar probe Lunar Prospector flew over this area at a low altitude and registered strong magnetic anomalies beneath the lunar surface (a magcon). The Descartes region obviously still holds many unsolved riddles.

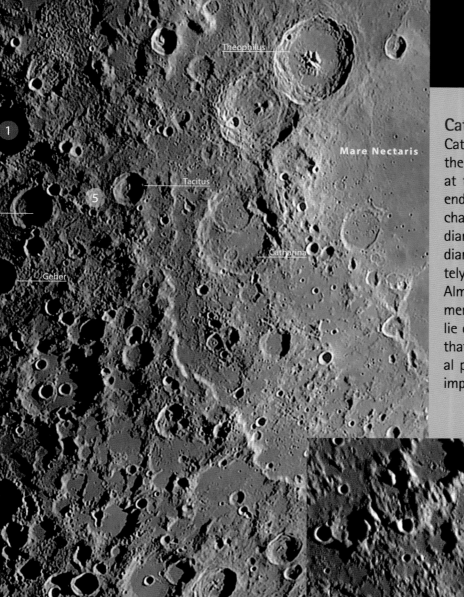

Theophilus

Mare Nectaris

Tacitus

5

Almanon

Geber

Catharina

1

3

Catena Abulfeda 17.0°S, 17.0°E (5)

Catena Abulfeda is, with a length of 210 km, by far the best-known crater chain on the Moon. It begins at the southeastern rim of Abulfeda (65 km) and ends at the northern edge of the Rupes Altai. The chain consists of about 20 individual craters with diameters between 1 and 3 km. Because of the small diameters of the craters, a large telescope is definitely required for its observation. The largest crater is Almanon C, roughly in the middle. Smaller instruments show only the largest craters. All the craters lie on a straight line. That leads to the supposition that the impactor was broken into many individual pieces by gravitational forces shortly before the impact.

Apollo 16

Andel

Dollond

Kant

Descartes A

4

1

5

1

5

Almanon C

Tacitus

Almanon

Geber

2

3

5

Theophilus 11.4°S, 26.4°E

An impressive complex crater with a diameter of 110 km, which was created in the Eratosthenian period on the lunar timescale. The crater wall reaches a height of 4.4 km above the crater's floor and the huge, triple central peaks attain a height of 1.4 km above the floor. The crest of the rim is about 1.2 km above the surrounding terrain. Except for the western rim, the complete inner wall is marked by prominent terraces. At high magnifications they appear as linear, sectional landslips. The northwestern wall is broken by the small crater Theophilus B (8 km). The central peak is strongly split and has low foothills on the southern side. The crater floor appears smooth and flat and is filled with melted impact material, which is also emplaced right round outside the crater rim. In addition, large telescopes show a few crater pits on the floor.

Shortly after First Quarter, Theophilus appears bright and shows the beginnings of a ray system, which becomes more prominent with increasing solar altitude. About 100 km southeast lies Theophilus L, a small, circular crater 5 km in diameter, that has a halo of very dark material. It is a Dark Halo Crater.

Mädler 11.0°S, 29.8°E

Mädler is a very interesting crater, 27 km in diameter, lying east of Theophilus. It is the source of a very broad, fan shaped deposit of ejected material, which implies a grazing impact (as with Proclus). The fan extends in an easterly direction.

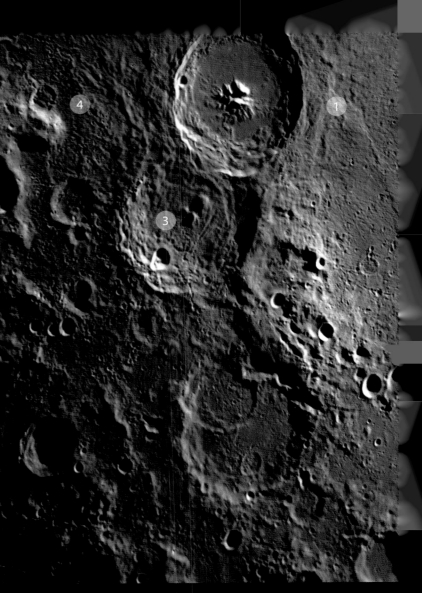

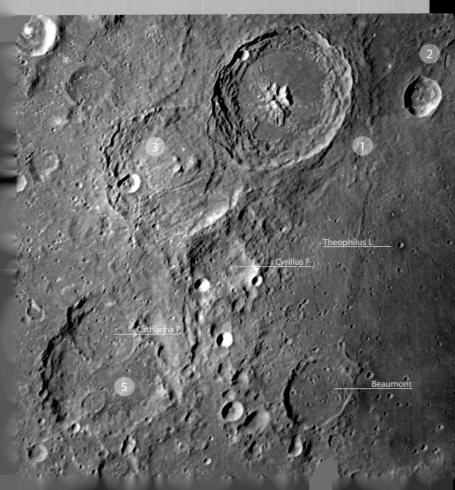

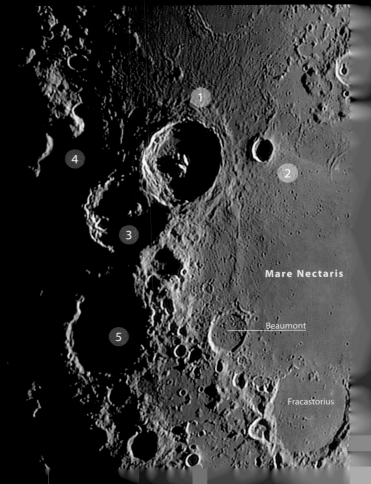

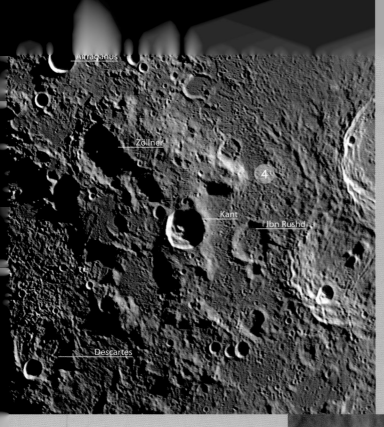

Cyrillus 13.2°S, 24.0°E 3

Cyrillus is also classed as a complex crater, and has a diameter of nearly 100 km. The crater floor shows three central peaks, eccentrically placed, of which two have a rounded shape. The crater walls of Cyrillus, in contrast to those of Theophilus, are heavily eroded, and it is perfectly clear that the Theophilus impact destroyed part of the northeastern wall of Cyrillus. This clearly implies that Cyrillus must be older than Theophilus. The southwestern wall is broken by the pear-shaped double crater Cyrillus A (17 km). The western wall of Cyrillus F (44 km, lying south of Cyrillus) and the terraces surrounding Cyrillus are extremely bright in sunlight. The crater floor appears rough, and furrowed, and is covered in ranges of hills and valley-like structures. The floor of Cyrillus F has a central elevation and has a slightly convex curvature, which when the Moon is about 19 days old, casts a shadow like the shell of a tortoise.

Mons Penck 10.0°S, 22.0°E 4

Mons Penck lies about 100 km west of the northern crater wall of Theophilus. It is an outlier of the southern highlands (known here as the Kant Plateau) and is about 30 km across, with a difference in height of about 4 km from the eastern plains. The gradient amounts to between 10° and 30°. When illuminated near the terminator it is an imposing sight.

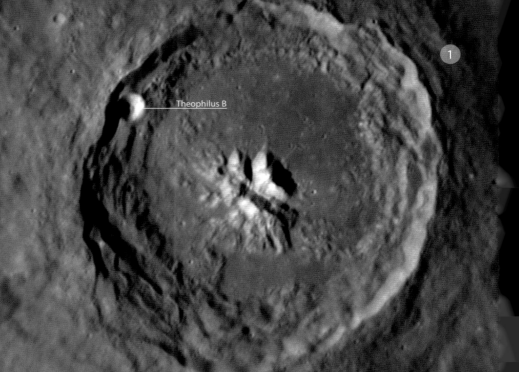

Catharina 18.1°S, 23.4°E 5

Catharina is a heavily damaged complex crater with a diameter of 104 km. The crater is linked to Cyrillus by a broad valley- or trough-like structure, through the centre of which runs a short, irregular chain of craterlets. Catharina is probably the oldest of the three craters and has been almost completely destroyed by later impacts. There are numerous smaller and larger crater on the floor, but no central peak is visible. Catharina P is the largest of the craters on the crater floor and, at 45 km, is almost half as large as Catharina itself. Given its diameter, Catharina must once have had a central peak, but this was probably completely destroyed by the impact of Catharina P.

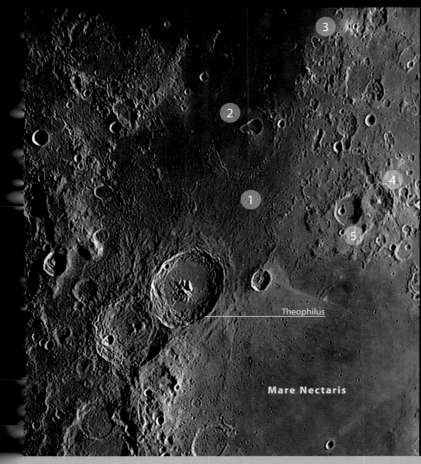

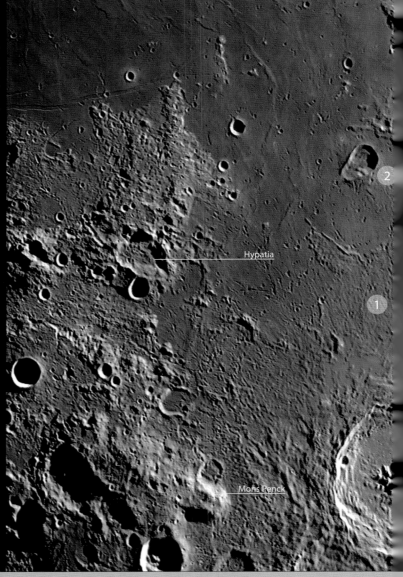

Theophilus

Mare Nectaris

Hypatia

Mons Penck

Sinus Asperitatis 6.0°S, 25.0°E

Sinus Asperitatis, 'Bay of Asperity', is a lava-covered surface about 180 km across. It lies north and east of the crater Theophilus and in the north passes into Mare Tranquillitatis. It has an appropriate name, because the surface of the lava is rough, furrowed, and uneven.

Torricelli 4.6°S, 28.5°E

Torricelli is a circular crater of about 22 km in diameter. The western crater wall is broken and overlapped by a smaller elliptical crater. Together both craters appear pear-shaped. Torricelli lies within a nameless, significantly larger ghost crater,

the remnants of the crater wall of which may be seen only in the southeastern direction. The rest has been submerged by the Sinus Asperitatis lava. The whole region has been covered by ejecta from the Theophilus impact and is correspondingly craggy.

Censorinus 0.4°S, 32.7°E

A small, funnel-shaped crater with a diameter of just 3 km, lying on the southern border of Mare Tranquillitatis. It is surrounded by a halo of very bright ejecta. At high solar elevations it appears as one of the brightest objects on the Moon's surface.

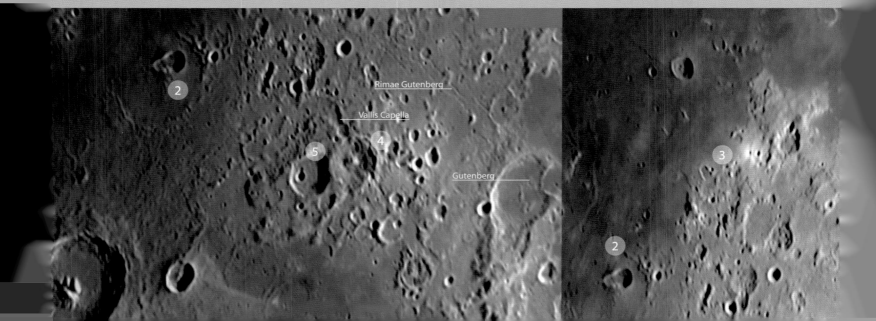

Rimae Gutenberg

Vallis Capella

Gutenberg

Maskelyne

Mare Tranquillitatis

③

②

①

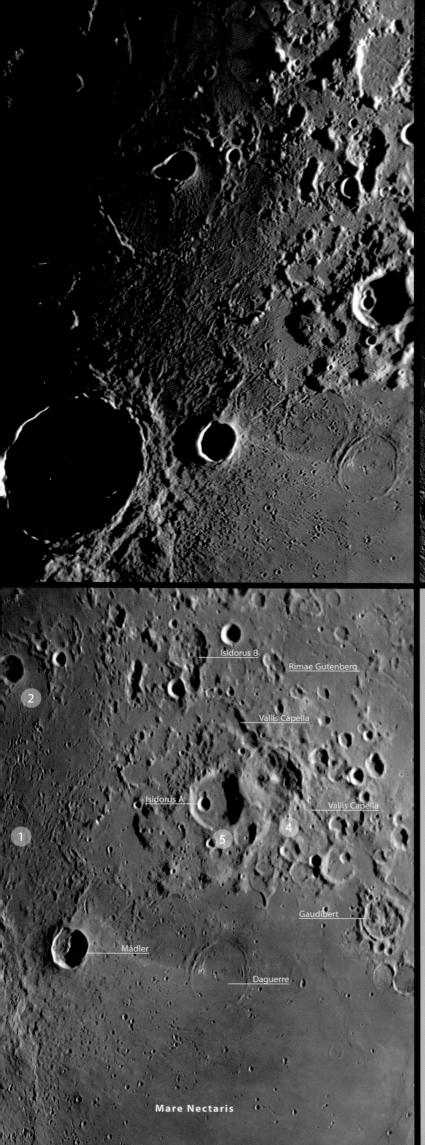

Isidorus B

Rimae Gutenberg

②

Vallis Capella

Isidorus A

Vallis Capella

④

①

⑤

Gaudibert

Mädler

Daguerre

Mare Nectaris

Capella 7.5°N, 35.0°E ④

A crater, 49 km in diameter, lying on the northern edge of Mare Nectaris. Its central peak is extremely large and wide relative to the crater's diameter, and has a depression at its summit. The central peak was probably altered by post-volcanic events. Vallis Capella, a valley about 110 km long, crosses the crater, running in a southeasterly direction and has a slightly tapering appearance. It consists of multiple overlapping, partially eroded, individual craters and is part of the Imbrium Sculpture. The valley was created by secondary impacts, triggered by the Imbrium impact.

Isidorus 8.0°S, 33.5°E ⑤

Isidorus is a crater with a diameter of 42 km and a very flat, smooth crater floor. It has no central peak and the eastern crater wall has been broken by the crater Capella. A small, prominent crater, Isidorus A (10 km), lies on the crater's floor near the western wall. An unnamed chain of overlapping craterlets runs near Isidorus B (30 km).

Statio Tranquillitatis

Statio Tranquillitatis 0.7°N, 23.5°E ①

On 20 July 1969, the landing module Eagle of the first manned Moon mission, Apollo 11, landed east of the crater Sabine. The landing site is a very flat, level region and was officially named Statio Tranquillitatis. The location was consciously chosen by the planning team, to minimize the dangers to the first landing from an uneven surface and craterlets. The landing approach was from the east, with the Sun behind them. To honour the first manned Moon landing, three small craters were officially named by the IAU after the astronauts: Armstrong (4.6 km), Collins (2.4 km), and Aldrin (3.4 km). Neil Armstrong was the first man to set foot on the Moon, followed by Buzz Aldrin. Collins remained in lunar orbit.

Rimae Hypatia 1.0°S, 23.0°E ②

A very broad, linear rille on the border between Mare Tranquillitatis and Sinus Asperitatis (the 'Bay of Asperity' or 'Bay of Roughness'), and which splits into a 'Y'-shape at its eastern end. The crater Moltke B lies right where it splits. The crater pit Moltke AC lies directly on top of the rille. The main segments of Rimae Hypatia are extremely broad and easy to observe even in small telescopes.

Moltke 0.6°S, 24.2°E ③

South of the Apollo-11 landing site, and north of the main portion of Rimae Hypatia lies the small, but conspicuous circular crater, Moltke (6 km in diameter). It is surrounded by a halo of bright ejecta and served as a navigation marker for Neil Armstrong during the Apollo mission's landing approach.

Rimae Ritter

Rimae Ritter

Sabine 1.4°N, 20.1°E 6
Ritter 2.0°N, 19.2°E 7

Two neighbouring craters, both of 30 km in diameter, lying near the lunar equator. A prominent mountain ridge breaks into the southern wall of Sabine. Sabine and Ritter belong to the class of FFC craters. Northwest of Ritter is the rille system Rimae Ritter (total length about 100 km) and southeast of Sabine are the Rimae Hypatia (total length about 200 km). Both are linear rille systems (tectonic fracture zones), lying on the edge of the mare and which follow the course of the boundary between the lava and the highlands.

Delambre 1.9°S,17.5°E 4

Delambre is a remarkably fine, symmetrical crater, 51 km in diameter. The crater walls reach a height of 3 km above the crater floor, which may be seen to have a complex structure, even in small instruments. It lies in the southern highlands, on the borders of Mare Tranquillitatis, south of the craters Sabine and Ritter.

Dionysius 2.8°N, 17.3°E 5

Under Full-Moon illumination, Dionysius is a very bright crater, 18 km in diameter, that is surrounded by a halo of bright material. The crater is remarkable in that the ray system consists of both bright and dark material. Careful observation shows a pattern of dark stripes within the bright halo, one streak of which crosses the crater Ritter and ends at the wall of Sabine. The dark stripes were first noticed in 1994 on images returned by the Clementine lunar probe. The dark material consists of mare lava, and the bright material is anorthositic basement rock and was brought to the surface from greater depths by the impact. This radial 'zebra-pattern' of stripes is unique on the nearside of the Moon.

Mare Serenitatis

Montes Taurus

Littrow

Sinus Amoris

Vitruvius

6

7

8

Carrel

10

11

2

1

Ritter

Sabine

Maskelyne

Apollo 11

Delambre

Sinus Asperitatis

Montes Pyrenaeus

Theophilus

Mare Nectaris

Mare Tranquillitatis ①
8.5°N, 31.5°E

Mare Tranquillitatis, the 'Sea of Tranquillity', is an area of about 420 000 km², slightly larger than Mare Serenitatis that lies to its north. Its size is comparable with the Black Sea on Earth. Like all the maria, Tranquillitatis is also the central, lava-flooded region of a considerably larger impact basin, and associated with a gravitational anomaly (a mascon).

The enormous lava expanse of Mare Tranquillitatis was the target of some unmanned lunar probes: Ranger 6, Ranger 8 (1965, hard landings) and Surveyor 5 (1967, soft landing). Ranger 6 was a failure, because the cameras could not be activated shortly before impact. In 1969 the first manned lunar landing with Apollo 11 took place in the southern portion of the mare. The landing site of the landing module Eagle lay near to Ranger 8 and Surveyor 5, east of the craters Ritter and Sabine.

Mare Tranquillitatis offers a whole range of extremely interesting lunar formations for both small and large telescopes. The graduated brightness of the lava boundary southwest of the crater Carrel is remarkable, as is the lava boundary between Mare Serenitatis and Mare Tranquillitatis, north of the crater Plinius; the formation Lamont in the southwestern portion of Tranquillitatis is unique on the Moon's nearside; the lunar domes near Arago; the region around Cauchy and Jansen; and the broad rille systems on the western border of Tranquillitatis: Rimae Hypatia, Ritter, Sosigenes, and Maclear.

Cauchy 9.6°N, 38.6°E ②

An extensive region around the crater Cauchy is one of the most interesting areas of the Moon for any size of telescope. Cauchy is a small crater, only 12 km across, with a sharp crater wall, which is clearly visible even at high Sun elevations. Southwest lies Rupes Cauchy, a linear fault, like an escarpment where the ground has dropped away, that is about 120 km long. Northeast lies Rima Cauchy, a broad, 210 km long, linear rille. (Both objects resemble Rupes Recta and Rima Birt.) It broadens out to 4 km in width and is thus also visible in small telescopes.

Rupes Cauchy is, in this respect, almost unique, in that the slope at the southeastern end (near the crater Cauchy C, 4 km) changes into a rille. (A similar transition of a fault to a rille may be observed with Rimae Bürg in Lacus Mortis.) At low solar elevations, the northeastern slope of Rupes Cauchy casts a conspicuous shadow, and at high elevations is really bright.

Rupes and Rima Cauchy lie on the outer northeastern periphery of Mare Tranquillitatis and are oriented radially to the centre of the Imbrium Basin. The trigger for their formation was probably the immense lava flows that had flooded the Tranquillitatis Basin, and their final form arose through the shock waves caused by the subsequent impact that led to the formation of the Imbrium Basin.

South of Rupes Cauchy two volcanic domes, Cauchy Tau and Cauchy Omega, with diameters of about 14 km and 10 km, respectively, are observable. Both domes are only about 200 m high, and so are visible only under very low solar illumination. The summit crater (the caldera) on dome Cauchy Omega has been given the name Donna, and has a diameter of about 2 km. Dome Cauchy Tau has a few, very difficult to observe, hollows on its downward slopes. Farther west lie other, but significantly smaller, domes.

Aryabhata 6.2°N, 35.1°E ③
Lyell 13.6°N, 40.6°E ④
Lucian 14.3°N, 36.7°E ⑤

Aryabhata and Lyell are the remnants of craters almost completely submerged by lava, 22 km and 32 km in diameter, respectively. Lyell has an extremely dark crater floor. Lucian is a sharply defined, small crater with a diameter of just 7 km. Near all three craters, volcanic domes may be observed under grazing illumination.

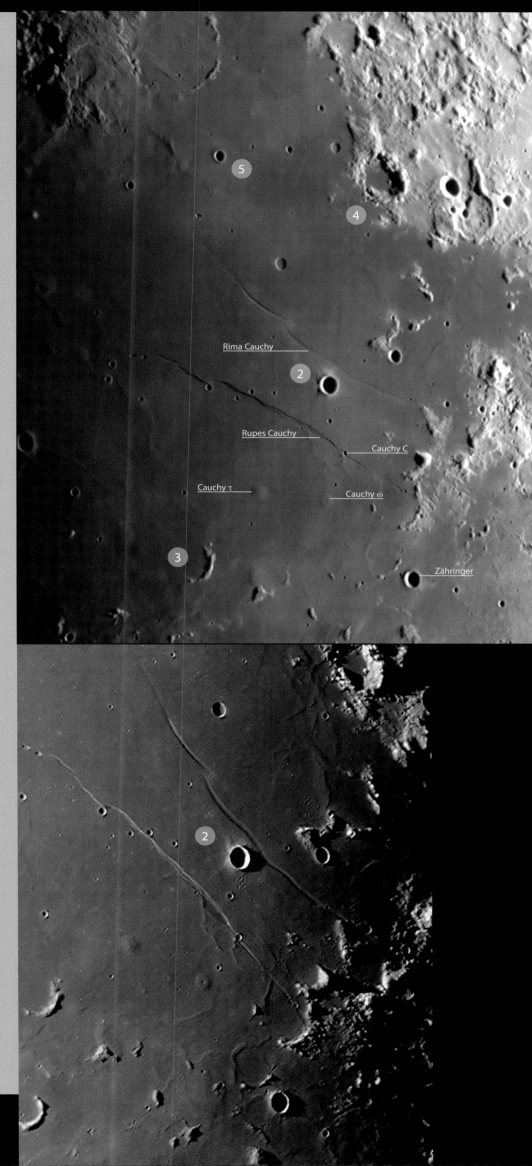

Rima Cauchy

Rupes Cauchy

Cauchy C

Cauchy τ

Cauchy ω

Zähringer

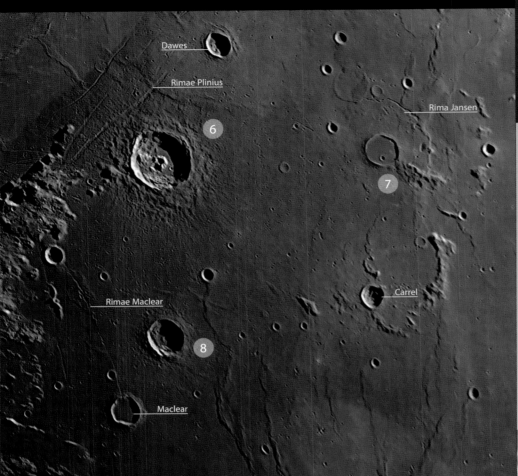

Plinius 15.4°N, 23.7°E **6**

A very conspicuous crater with a diameter of 43 km, a central peak and terraced inner walls. The central peak, which has partially destroyed slopes that are clearly visible under low illumination, exhibits a summit crater, and directly alongside lies a crater pit. A system of parallel, linear rilles – the Rimae Plinius, with an overall length of about 124 km – lies to the north of Plinius and intersect some of the mare wrinkle ridges.

The crater is named after Gaius Plinius Secundus (23 to 79 AD). Pliny was the author of a 37-volume encyclopaedia, entitled 'Naturalis Historia'. He died during the eruption of Vesuvius that destroyed Pompeii.

Jansen 13.5°N, 28.7°E **7**

A lava-flooded crater with flat walls and a diameter of 23 km, not to be confused with the large complex crater Janssen in the southern highlands at 45.4°S, 40.3°E. Northeast of Jansen lies the sinuous rille Rima Jansen with a length of about 35 km. In the nearby surroundings a few other submerged ghost craters may be observed.

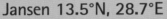

Ross 11.7°N, 21.7°E **8**

A crater that is 24 km in diameter, not to be confused with the small crater Rosse in Mare Nectaris.

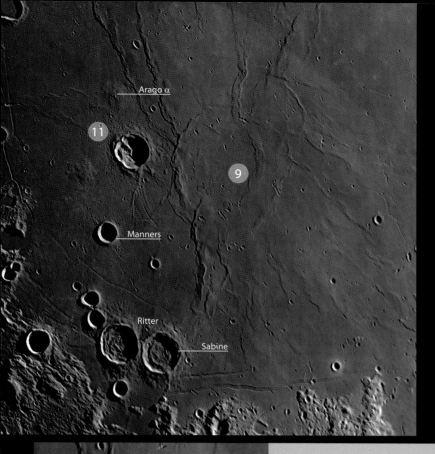

Lamont ⑨
4.4°N, 23.7°E

Lamont is a unique feature on the nearside of the lunar surface. It is probably a ghost crater completely submerged by lava, surrounded by two concentric, mare wrinkle-ridge systems, one of which shows radial structures. Lamont is the centre of a gravitational anomaly (a mascon). As a feature, Lamont is very flat, so that observation requires extreme grazing illumination.

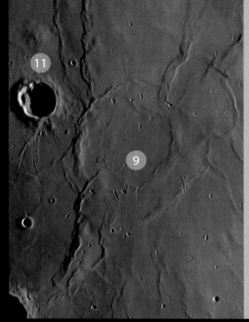

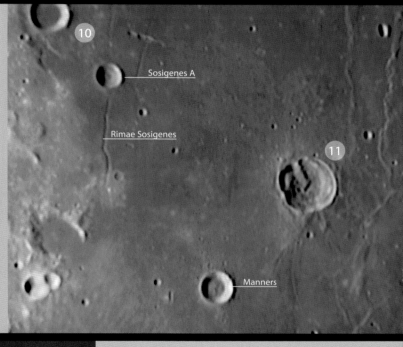

Sosigenes ⑩
8.7°N, 17.6°E

Sosigenes is a small crater, 18 km in diameter. It has a small central peak and completely smooth, level crater floor. To the southeast lies the small, circular crater Sosigenes A (c. 9 km). Further south lies Rimae Sosigenes, a linear rille system, consisting of several rilles running parallel to one another with a length of 150 km. Towards the northwest there are valley-like structures of the Imbrium Sculpture, and Sosigenes itself was probably created by a secondary impact.

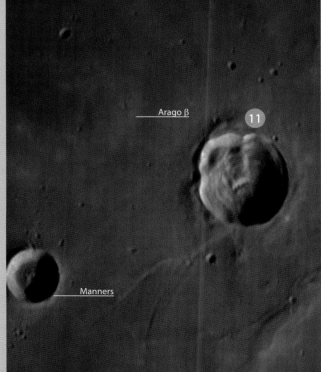

Arago 6.2°N, 21.4°E ⑪

A crater, 26 km in diameter, in the western portion of Mare Tranquillitatis. It has sharply defined crater walls and a mountainous crater floor. North lie the lunar domes Arago Alpha and (to the west) Arago Beta. Both domes are relatively large and may be observed with small telescopes when illuminated near the terminator. In large telescopes, some details become visible on the flanks of the domes. The shape of Arago Alpha and Beta resembles that of large Icelandic shield volcanoes. Farther north of Arago Alpha – more-or-less in the middle between the craters Arago and Ross – there is another group of four smaller volcanic domes.

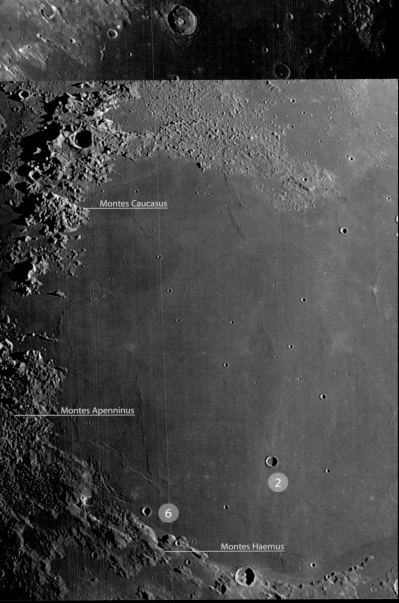

Lacus Mortis

Lacus Somniorum

Posidonius

③

① ⑤ ⑨

②

⑥

Menelaus

Montes Caucasus

Montes Apenninus

Montes Haemus

② ⑥

Mare Serenitatis ❶
28.0°N, 17.5°E

Mare Serenitatis, the 'Sea of Serenity', is a large, nearly round lava expanse with an area of about 500 × 700 km. Like all the other maria, Mare Serenitatis is also the centre of a gravitational anomaly (a mascon), and it is also a typical multi-ring basin with at least two basin ramparts. The darker edge regions suggest that this lava is richer in metallic components such as iron and titanium, and that it emerged earlier in the period of lava flows than the lava in the central region. Around the edge, the lava is also thinner than in the centre of the mare.

From soil samples obtained during the Apollo 17 mission (the landing site was on the southeastern edge of the mare), geologists were able to determine that the lava in Mare Serenitatis erupted at different times, almost 800 million years apart (between 3.8 and 3.0 billion years ago). Even on the Earth and for terrestrial geological sequences, this is an enormous time-span. This period was the most active one for lunar volcanism. About 3 billion years ago, the lava flows gradually ceased, but did not stop completely. The last active phase of lava production took place about 1 billion years ago.

The southwestern edge of Mare Serenitatis is bordered by Montes Haemus, a range of mountains that is about 400 km long. The range is the visible remnant of the wall of the Serenitatis Basin.

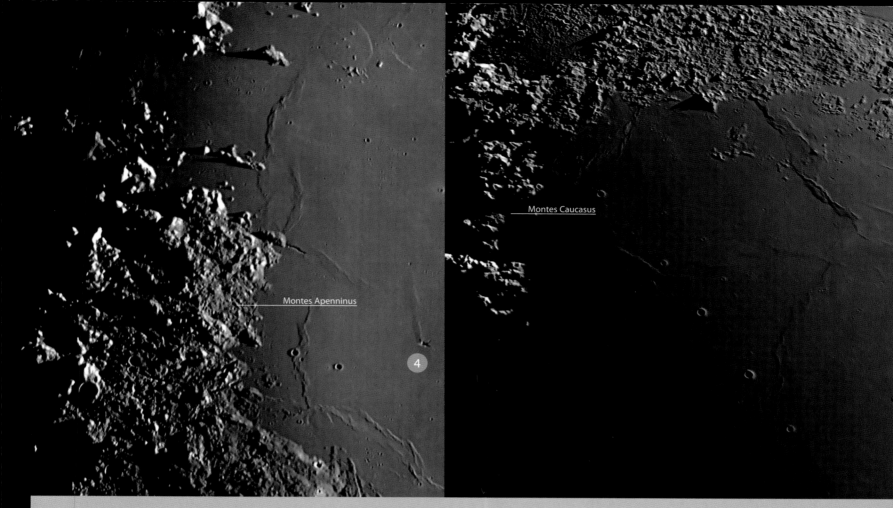

Montes Caucasus

Montes Apenninus

4

Bessel 21.8°N, 17.9°E   ②

A small, prominent crater in Mare Serenitatis, with a diameter of 15 km. It lies on a wrinkle ridge, and its western portion is covered by a bright ray of ejecta from the crater Menelaus.

Linné 27.7°N, 11.8°E   ③

Linné is a small, funnel-shaped crater of just 2.4 km in diameter, which is surrounded by a halo of bright ejected material. Under high solar illumination it appears as a white spot on the dark lava surface of Mare Serenitatis. In the second half of the 19th century many publications appeared, reporting changes and even the disappearance of the crater. The origins of these reports were undoubtedly observations made under poor seeing conditions or observations with inadequate optics.

Rima Sung-Mei 24.6°N, 11.3°E   ④

Rima Sung-Mei is a very narrow rille, only 4 km long, that ends in a sort of depression. It lies in the immediate neighbourhood of Dorsum Owen and is only detectable visually with apertures over 10″ (250 mm) at the highest magnifications and under extremely good seeing conditions.

Two valley-like sumps branch out of the depression. They are called Vallis Christel and Vallis Krishna. The names were accepted into lunar nomenclature by the IAU after the flight of Apollo 15. They are probably the result of multiple volcanic events. Sung-Mei is a Chinese female name.

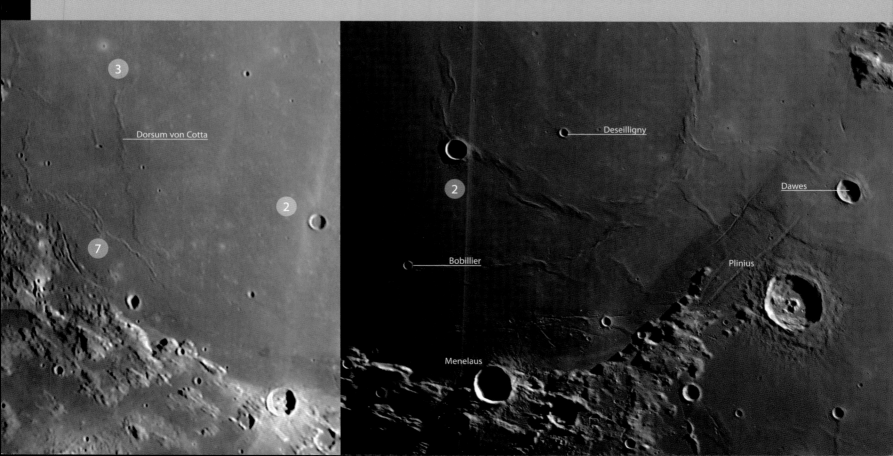

Dorsum von Cotta

Deseilligny

Dawes

Bobillier

Plinius

Menelaus

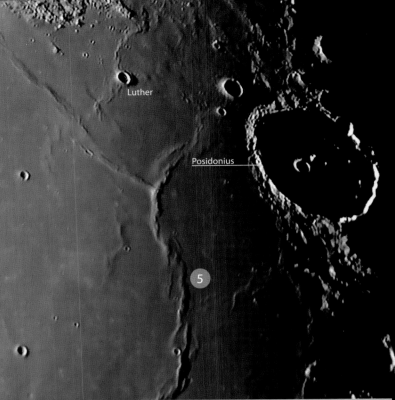

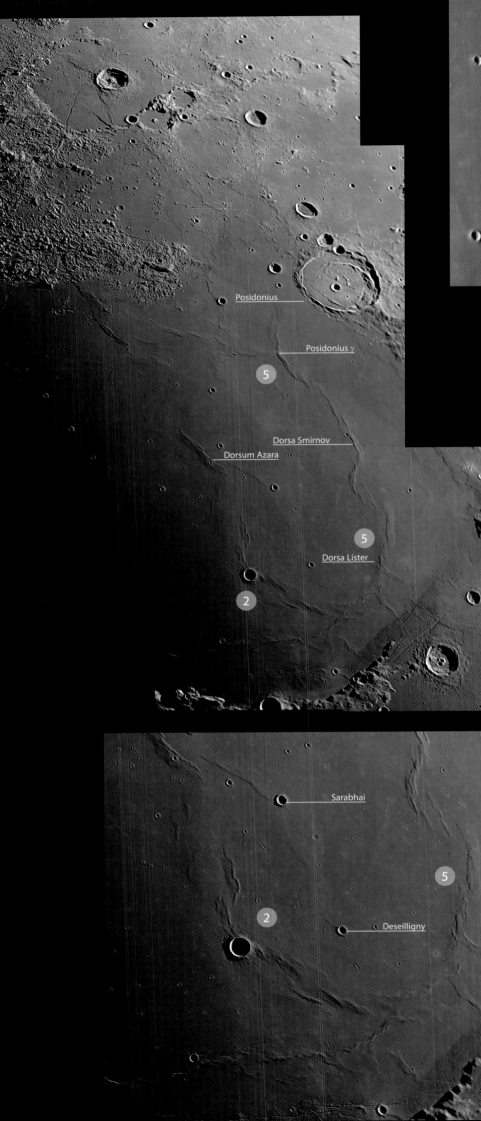

Serpentine Ridge ⑤

The eastern side of the mare surface is crossed by a series of low mare wrinkle ridges. Unofficially these are known as the Serpentine Ridge. The wrinkle ridges consist of two main segments, which are officially known as Dorsa Lister and Dorsa Smirnov. Both are about 10 km wide and extend, more-or-less following the trend of the edge of the mare, over considerable distances: Dorsa Lister for around 300 km and Dorsa Smirnov for about 130 km.

South of Dorsa Lister lies the crater Plinius with a diameter of 43 km. In between, oriented east–west is the Rimae Plinius rille system, with a length of 124 km. The small crater Very, with a diameter of just 5 km lies almost in the centre, directly on top of the ridge of Dorsa Smirnov.

Posidonius Gamma is the designation, no longer used, for a small peak on the Dorsa Smirnov wrinkle ridge. The mountain has a summit crater, which is surrounded by a small halo of bright ejecta. The rilles (Rimae Littrow, Rimae Chacornac) on the eastern edge of Mare Serenitatis are fracture zones, created by a subsidence of the Serenitatis Basin's central lava area.

In the western portion there are Dorsum Buckland, Dorsum Azara and a few others. All these mare wrinkle ridges in Mare Serenitatis branch and partially intersect one another. They are, in general, no more than about 200 m high, and are therefore only visible at very low solar elevations, when near the terminator.

Sulpicius Gallus 19.6°N, 11.6°E

Rimae Sulpicius Gallus 21.0°N, 10.0°E

Sulpicius Gallus is a small, round crater, 12 km in diameter with a sharp crater wall. Adjoining it to the northwest are the Rimae Sulpicius Gallus. This is a system of linear rilles, that extends for a length of 90 km parallel to the Montes Haemus mountain range. They cross the Sulpicius Gallus Formation (a Dark Mantle Deposit). They are tectonic fracture zones, created by tension movements in the layers of lava at the edge of the mare, when the tremendous lava flows sank after they cooled and with the cessation of the pressure from the Moon's interior.

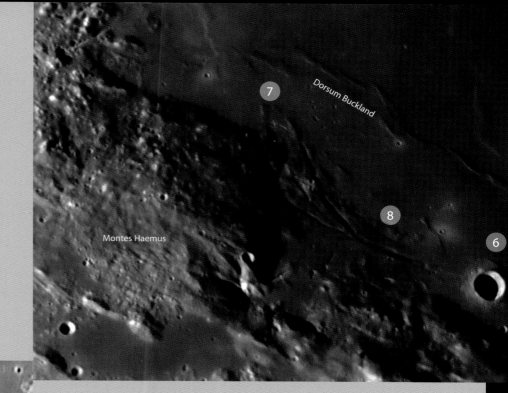

Sulpicius Gallus Formation 20.0°N, 10.0°E

A region at the southwestern edge of Mare Serenitatis has been called the Sulpicius Gallus Formation, and lies between the craters Sulpicius Gallus and Joy, with is only 6 km across. It stands out because of its dark coloration.

The Sulpicius Gallus Formation is undoubtedly a Dark Mantle Deposit (DMD) region, created by intense, explosive, pyroclastic volcanism, which has covered the surrounding region with dark, vitreous ashes. This active volcanism has been dated to long after the mare was flooded with lava. In the literature, other DMD regions that lie nearby (Mare Vaporum and Sinus Aestuum) are often included under the generic term as Sulpicius Gallus regions.

Le Monnier 26.6°N, 30.6°E

A lava-flooded crater, 60 km in diameter, with a very dark floor. It forms a bay of Mare Serenitatis (appearing similar to the crater Letronne, north of Gassendi). The crater floor is flat and level and, in medium-sized telescopes, shows no sign of craterlets or pits. In larger instruments, however, under high solar illumination it exhibits a few bright spots, which prove to be pits with bright haloes. Le Monnier was the target for the Luna 21 Moon probe, which carried the Lunakhod 2 lunar rover.

Luna 21 landed on 15 January 1973 on the southern edge of the 60 km wide crater Le Monnier, in the transition zone between Mare Serenitatis and the highlands in the Taurus mountains. The crucial factor in the choice of landing site was a tectonic graben that extends here in a north-south direction for just 16 km. The width of the rille averages 300 m, with a depth that reaches 80 m. During its five-month research traverse Lunakhod 2 travelled a total distance of 37 km. After 5 months the solar cells were covered in lunar dust, so the power supply to the on-board electronics failed, and the mission came to an end. Lunakhod did, however, carry retroreflectors and is still today regularly used for lunar laser ranging to determine the exact distance between Earth and the Moon.

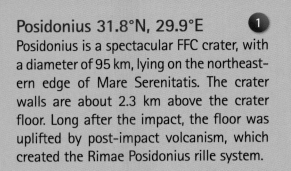

Posidonius 31.8°N, 29.9°E

Posidonius is a spectacular FFC crater, with a diameter of 95 km, lying on the northeastern edge of Mare Serenitatis. The crater walls are about 2.3 km above the crater floor. Long after the impact, the floor was uplifted by post-impact volcanism, which created the Rimae Posidonius rille system.

One of the rilles runs directly north at the foot of the western crater wall. At the end, the rille seems to run out into a series of crater pits. Along the eastern inner wall of the crater, following its curvature, there lies a prominent ridge (possibly landslides from the inner wall). Near the centre of the crater's floor is the crater Posidonius A (c. 15 km). Posidonius exhibits many remnants of central peaks. Northeast of the crater Posidonius A, separated by a broad rille, lies a central mountain with two sharp peaks. When observing with larger telescopes, they may be used to estimate seeing conditions and the resolution of the optics. The two peaks have a separation of 1.9 km which, at the Moon's average distance, corresponds fairly exactly to 1 arcsecond.

Posidonius A

Rimae Posidonius

Chacornac 2
29.8°N, 31.7°E

An almost completely destroyed crater, 51 km in diameter, directly adjoining Posidonius to the south. The crater rim has a slightly hexagonal shape. Rimae Chacornac, some small linear rilles on the crater's floor and south of the crater, are visible at high magnifications.

Daniell 3
35.3°N, 31.1°E

An elliptical crater with measurements of 23 × 29 km. It lies on the eastern edge of Lacus Somniorum. The linear rille system Rimae Daniell (37.0°N, 26.0°E) lie west of the crater, with an overall length of about 200 km and a maximum width of 3 km.

Rima G. Bond 33.3°N, 35.5°E 4

A fracture zone, just 170 km long and 4 km wide at its greatest. It runs in a straight line from the Taurus mountains to Lacus Somniorum and is visible even in smaller telescopes under appropriate lighting.

Hall 33.7°N, 37.0°E 5

A ruined, lava flooded crater, with a diameter of 35 km and an overall crescent shape. The rille system Rimae G. Bond lies directly west of the crater.

Lacus Somniorum

Eudoxus

Alexander

Mare Serenitatis

Lacus Mortis 45.0°N, 27.0°E

The 'Sea of Death' fills an extensively eroded large complex crater of about 150 km diameter, of which, fundamentally, only the western crater wall remains. The lava covers an area of c. 210 000 km² and lies to the south of the eastern end of Mare Frigoris.

Mason 42.6°N, 30.5°E
Plana 42.2°N, 28.2°E

Lying south of Bürg is the overlapping pair of craters Mason and Plana (oriented east-west). Both craters are lava-flooded. Mason, with dimensions of 33 × 43 km, is a significantly elliptical crater. Plana has a central peak and a diameter of 44 km. Both craters seem partially ruined, and on both crater floors craterlets are visible in large telescopes.

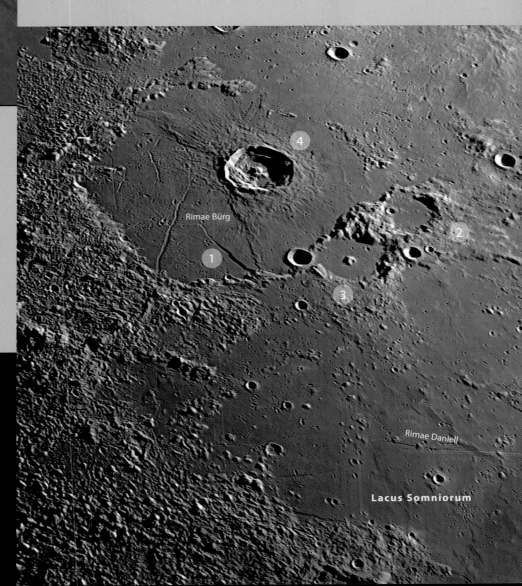

Rimae Bürg

Rimae Daniell

Lacus Somniorum

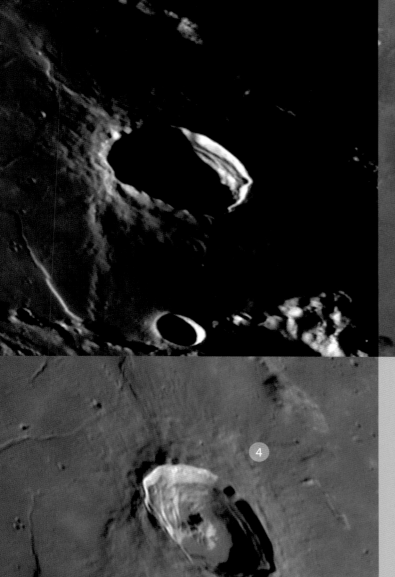

Bürg 45.0°N, 28.2°E   ④

Bürg is a prominent young crater (diameter 39 km) with a sharp-edged crater wall, that towers above the crater floor by about 2.2 km. The crater lies in the middle of a lava flow that was produced significantly earlier. Bürg has been dated to the Copernican period on the lunar timescale and in certainly one of the youngest of the larger craters on the nearside.

Bürg has a dominant central peak and terraced inner crater walls with many deep cracks and breaks. The ejecta from the impact lies in two masses to the north and – in a broad fan – to the south of the crater (somewhat similar to a mare wrinkle ridge). West and south of the craters there are two sections of a rille system (lying approximately at right angles to one another), the Rimae Bürg. The southern section changes from a rille to an escarpment (similar to Rupes Cauchy), while the western section is more like a fracture zone. On high-resolution Lunar Orbiter images, a few very small rilles and chains of overlapping craters are also visible northwest of Bürg.

Aristoteles 50.2°N, 17.4°E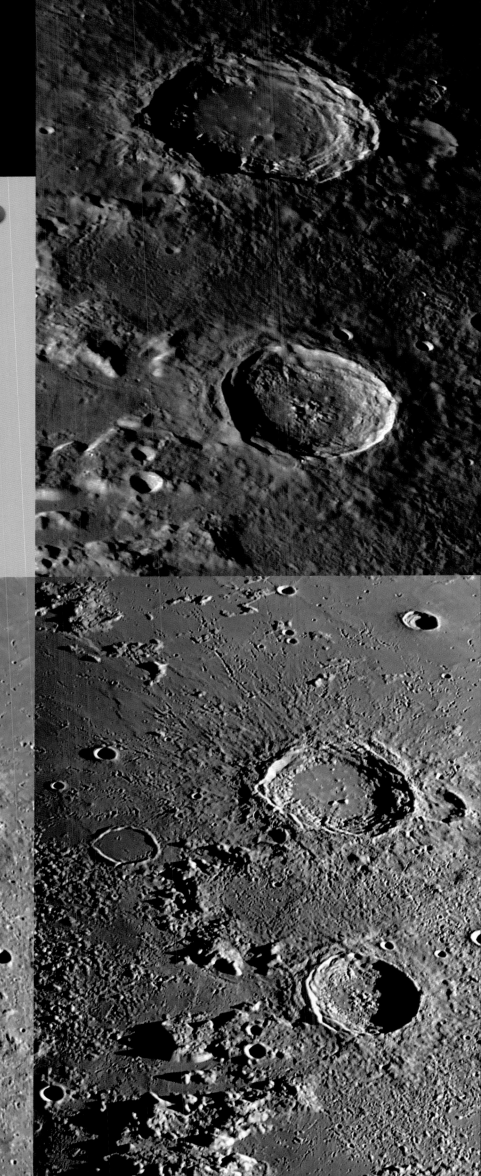

A very prominent, young, large crater with a diameter of 87 km, which was probably created during the Eratosthenian period on the lunar timescale. Aristoteles has pronounced terraced inner crater walls, and the crater floor lies 3.3 km below the crater rim. Remnants of central mountains are detectable. The crater floor is relatively smooth and is divided, in a north-south direction by a flat ridge (only visible under grazing illumination). Beneath the western wall of the crater, a landslide is clearly visible. A small crater chain lies on the southern wall and the eastern wall has been overlain by Mitchell (49.7°N, 20.2°E), a crater 30 km in diameter. Mitchell's crater floor is slightly convex. As with the crater Aristillus, Aristoteles is also surrounded by a broad, circular zone of melted impact material.

Galle

Mare Frigoris

Mitchell

Lamech

Montes Caucasus

Eudoxus 44.3°N, 16.3°E

Eudoxus is another prominent crater with terraced crater walls. It has a diameter of 67 km and a depth of 3.4 km. The floor of the crater, when compared with that of Aristoteles, appear significantly rougher. Both craters form a very conspicuous pair in the northeastern quadrant of the Moon.

Egede 48.7°N, 10.6°E

A crater, 37 km in diameter, that is almost completely filled with lava, with a somewhat angular shape, lying west of Aristoteles and Eudoxus. On the middle of the southern wall there is a crater pit, as there is on the northwestern portion of the crater floor. Large telescopes reveal about a dozen pits (between about 1 and 2 km in diameter) on the southwestern crater floor.

Montes Caucasus 1
38.4°N, 10.0°E
The Caucasus is a ruggedly craggy, heavily eroded, extensive mountain range, and the direct continuation of the Apennines, separated from the latter by an approximately 50 km wide, flat, lava-flooded 'strait'. The range stretches over a total length of approximately 520 km, and contains the highest peaks on the nearside of the Moon, which tower more than 6 km above the lava surface. To the east lies Mare Serenitatis, and to the west, Mare Imbrium. If one were a tourist standing on one of the highest mountains, one would be able to see for about 140 km.

Calippus 38.9°N, 10.7°E 2
A 32-km crater in the Caucusus highlands. The eastern crater wall runs almost in a straight line in the north-south direction. The western wall exhibits massive landslides. Very large instruments show a few crater pits on the crater floor. Calippus C, lying to the west, is a semicircular crater, 40 km in diameter, open to Mare Imbrium.

Rima Calippus 37.0°N, 13.0°E 3
A fairly wide, slightly curved rille, 40 km long, southeast of Calippus. It lies on the northwestern edge of Mare Serenitatis.

Alexander 40.3°N, 13.5°E 4
Alexander is a very heavily eroded complex crater, 81 km in diameter. The northeastern wall of the crater has been almost completely destroyed.

Valentine Dome 31.0°N, 10.3°E

At the southern end of the Caucasus and lying to the east, is a large, almost circular plateau, which is interpreted as being a volcanic megadome. Officially, the feature remains unnamed, but in lunar literature has the nickname 'Valentine Dome'. The plateau measures about 30 km in diameter, but is only about 100 m high, and may be observed only under extreme grazing illumination. Large telescopes reveal a few prominent cliffs on the plateau, together with crater pits and a narrow rille that runs eastwards, leaving the plateau and ending on the floor of Mare Serenitatis. The name was introduced by Alika Herring, a Hawaiian lunar observer, in 1962.

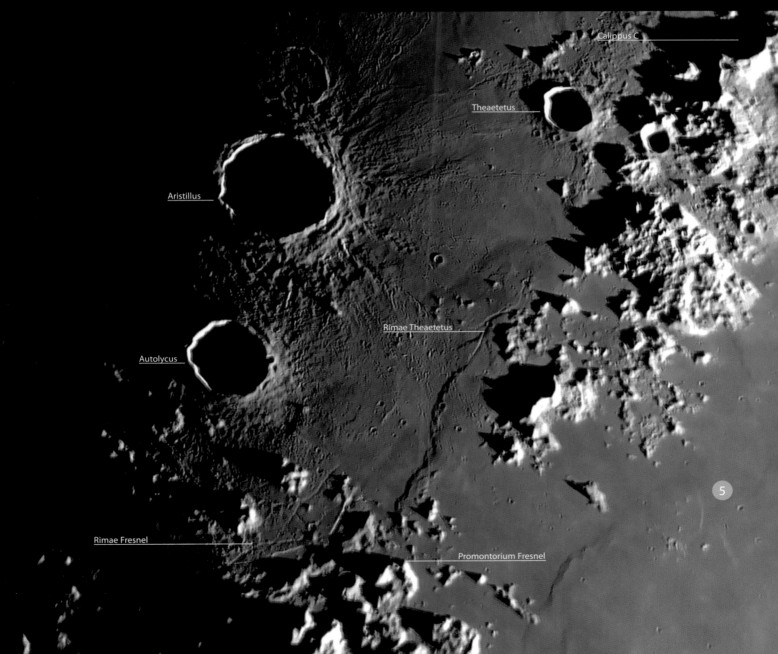

25 Autolycus/Aristillus

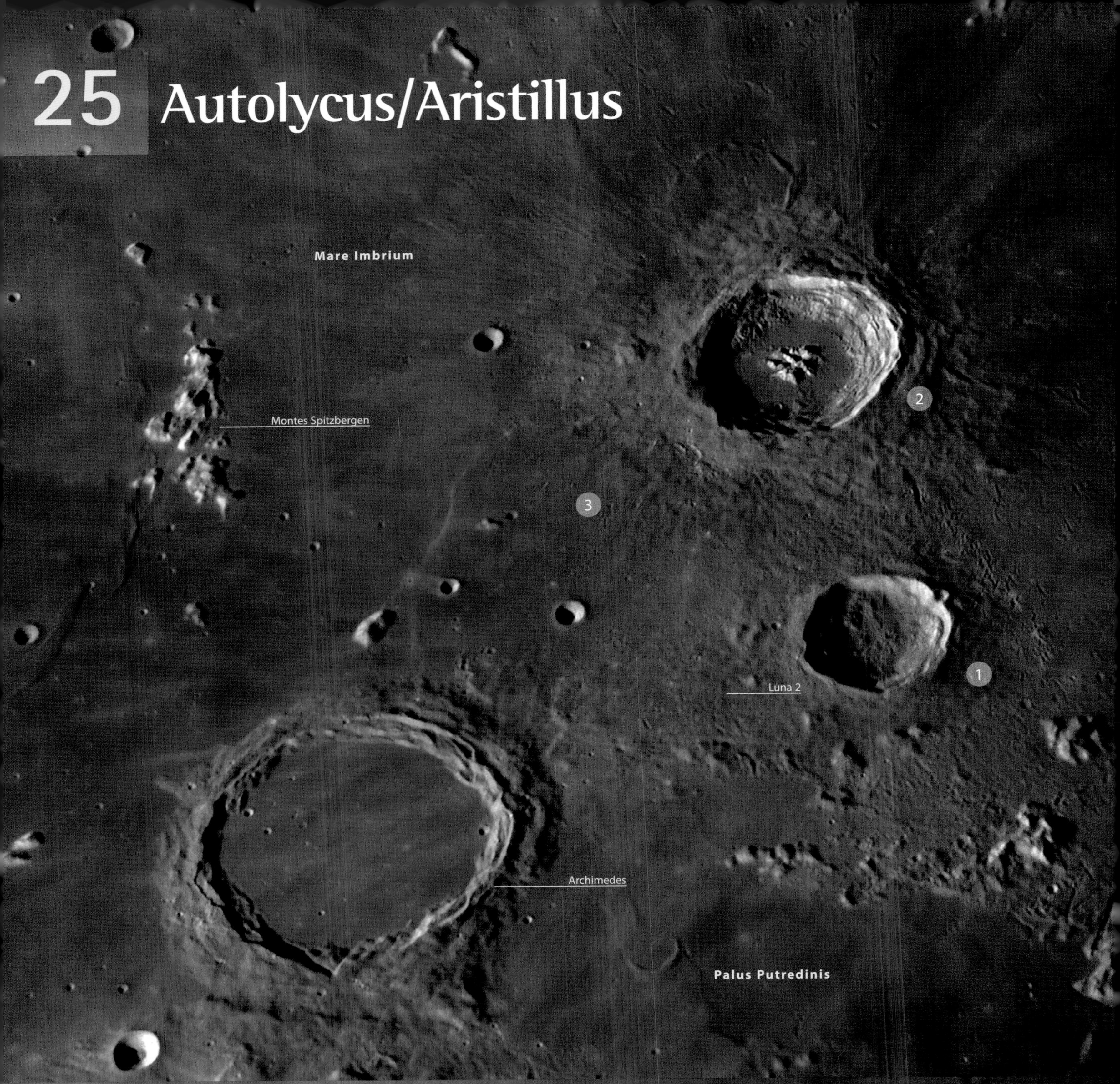
Mare Imbrium
Montes Spitzbergen
2
3
1
Luna 2
Archimedes
Palus Putredinis

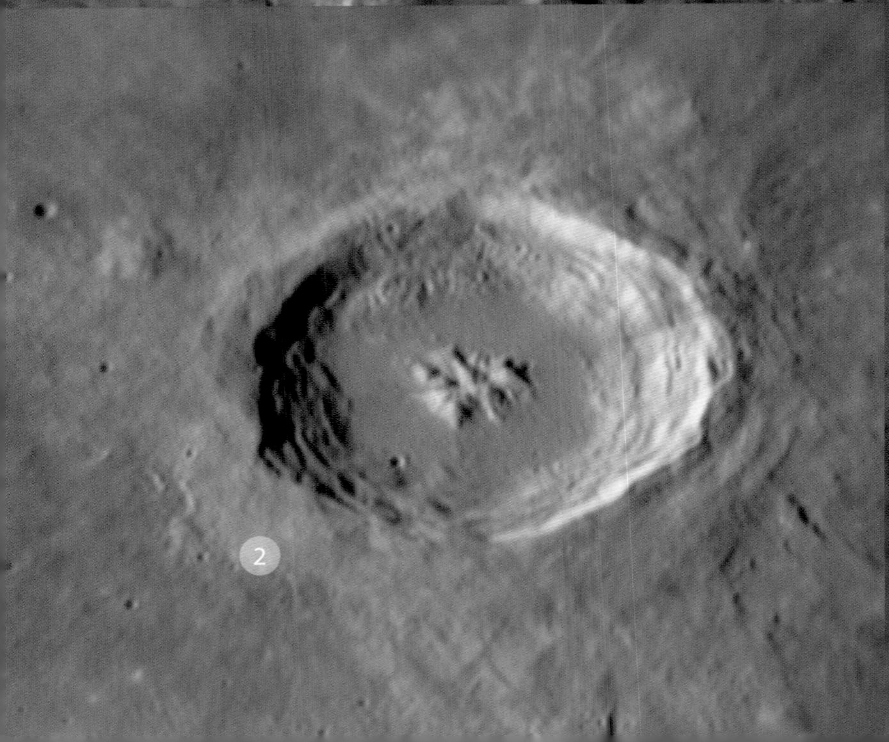
2

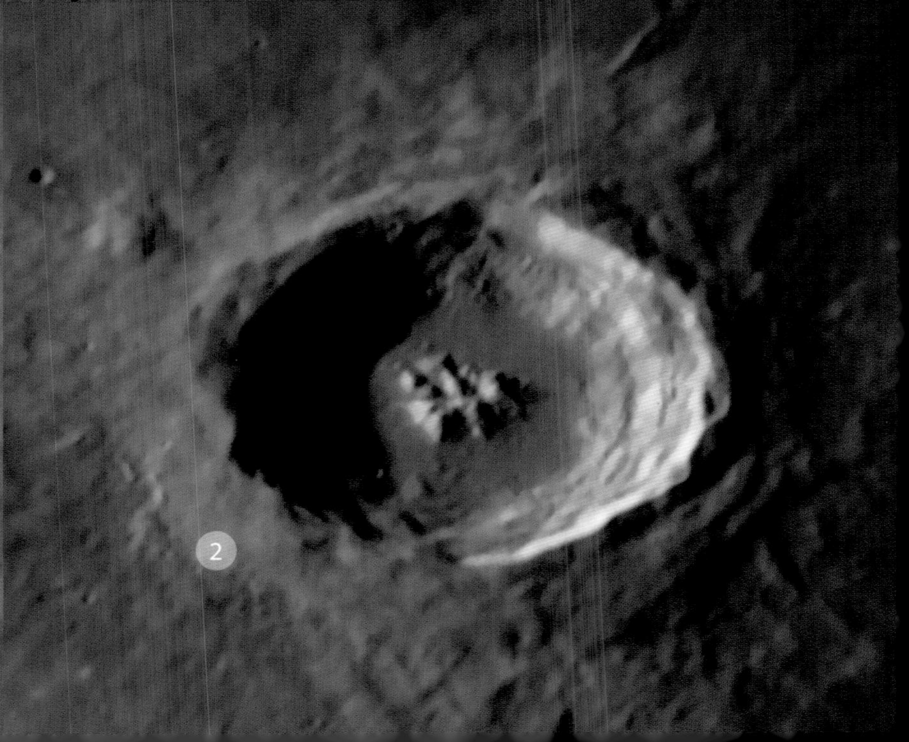
2

Mare Imbrium

Montes Caucasus

Mare Serenitatis

Montes Apenninus

Sinus Lunicus ③
32.0°N, 1.0°W

The Sinus Lunicus, 'Bay of the Moon', is a lava plain enclosed by two low ridges and lies between the craters Autolycus and Aristillus in the northeast and Archimedes in the southwest. In 1959, the Russian lunar probe Luna 2 made a hard landing on the lunar surface west of Autolycus. The region was officially named Sinus Lunicus by the IAU in 1970. Luna 2 was the first lunar probe to reach the Moon's surface. Its predecessor, Luna 1 missed the Moon by about 6000 km on 4 January 1959.

Mons Piton

Cassini

Piton γ

Theaetetus

Autolycus 30.7°N, 1.5°E ①
Aristillus 33.9°N, 1.2°E ②

A very conspicuous and prominent pair of craters in the Apennine region of Mare Imbrium. Under high solar illumination both craters exhibit ray systems. Autolycus, with a diameter of 39 km, has a very rough crater floor with a heavily eroded central peak. Aristillus has a diameter of 55 km. The inner walls of the crater are terraced and the central peak appears prominent above a level, smooth crater floor. The individual summits of the central mountain reach a height of up to 900 m above the floor. Particularly conspicuous is the broad annular zone of melted ejecta that is piled up outside the crater's wall. Directly north of Aristillus, this ejecta covers a submerged ghost crater, about 40 km in diameter, which may be seen only under grazing illumination.

Rimae Theaetetus

Archimedes

Rimae Fresnel

Cassini 40.2°N, 4.6°E

Cassini is a very conspicuous crater, 56 km in diameter, lying at the southern end of the Alps. It is a prominent landmark in this area of the Moon. The outer crater wall appears rounded and smooth because of deposits of melted ejected material – as with Archimedes and Aristoteles.

On high-resolution images the ejecta, which lies in a ring around the outer crater rim, appears remarkably smooth and wavy. It gives that impression that the impact onto the lunar surface took place into semi-fluid material with a consistency like that of mud. Considered morphologically, the crater and its immediate surroundings resembles those craters on Mars that are classified as 'rampart craters'. There, however, sub-surface ice plays a part, which is liquified by the impact and turns the soil surrounding the impact into a muddy consistency.

The floor of Cassini is slightly rounded and appears extremely smooth. A chain of hills stretches from the inner crater wall to the crater Cassini A (15 km). It is crossed by a narrow, valley-like structure. Cassini B (9 km) also lies inside the crater, and between them lie two mountain peaks. A craterlet of about 2 km in diameter lies directly on the northern crater wall.

Cassini M (8 km) lies immediately next to the outer rampart on the northwest. From there a valley-like trough runs more than 20 km northwards. An interrupted, very narrow, nameless rille lies between Cassini and the crater Calippus. Large telescopes using webcam techniques have revealed half-a-dozen additional crater pits on the floor of Cassini.

Running radially to the centre of Aristillus are crater grooves and pits created by secondary impacts following the main Aristillus impact, and which reach as far as the outer wall of Cassini.

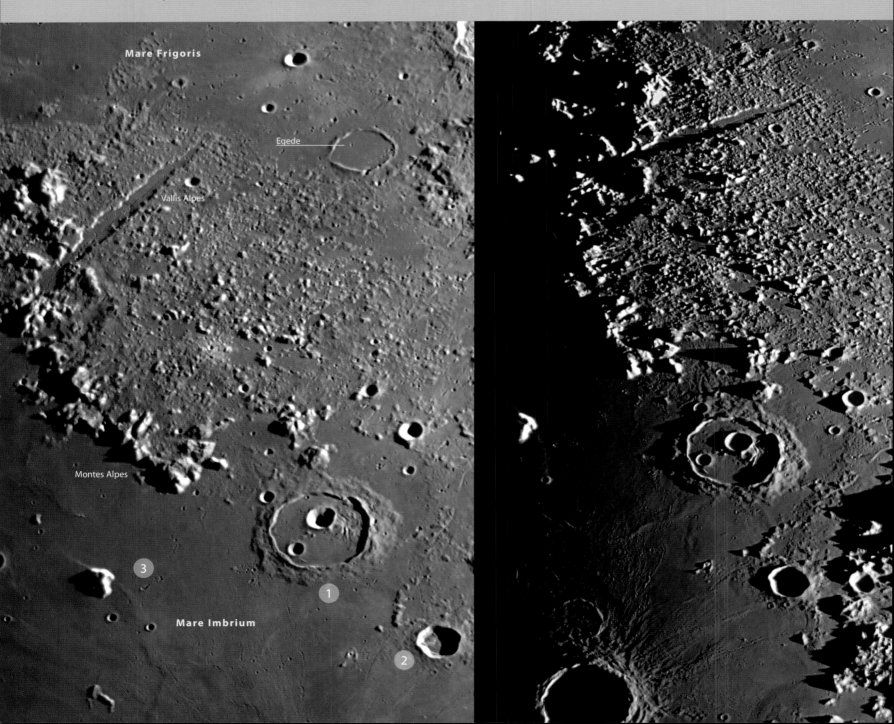

Mare Frigoris

Egede

Vallis Alpes

Montes Alpes

Mare Imbrium

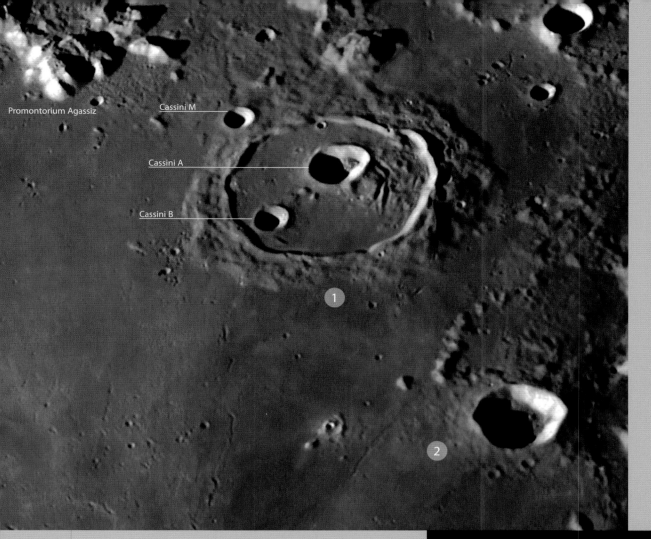

Promontorium Agassiz
Cassini M
Cassini A
Cassini B

1
2

Theaetetus
37.0°N, 6.0°E

A young crater with a sharply defined crater wall and a smooth floor, lying at the western foot of the Caucasus mountains. Its diameter is 25 km, and the wall of the crater is 2.8 km above the crater's floor. South of Theaetetus and about 100 km east of Autolycus and Aristillus, a linear rille system, Rimae Theaetetus (33.0°N, 6.0°E) runs south for a distance of about 50 km.

Mons Piton 40.6°N, 1.1°W

An isolated mountain massif with a base area of about 200 km² and a height of 2.3 km. Under grazing illumination it casts a long shadow on the lava of Mare Imbrium. Observations of Mons Piton when it is near the terminator (and casting its long shadow) give the impression that it is an extremely steep, needle-like peak. But if its height of 2.3 km is set against its area of about 200 km², it is obvious that Mons Piton is more like a high hill than a steep mountain. The gradient of the slopes is relatively low and the long shadow exceeds the height of the peak by a factor of 40.

Promontorium Deville

Promontorium Agassiz

Piazzi Smyth

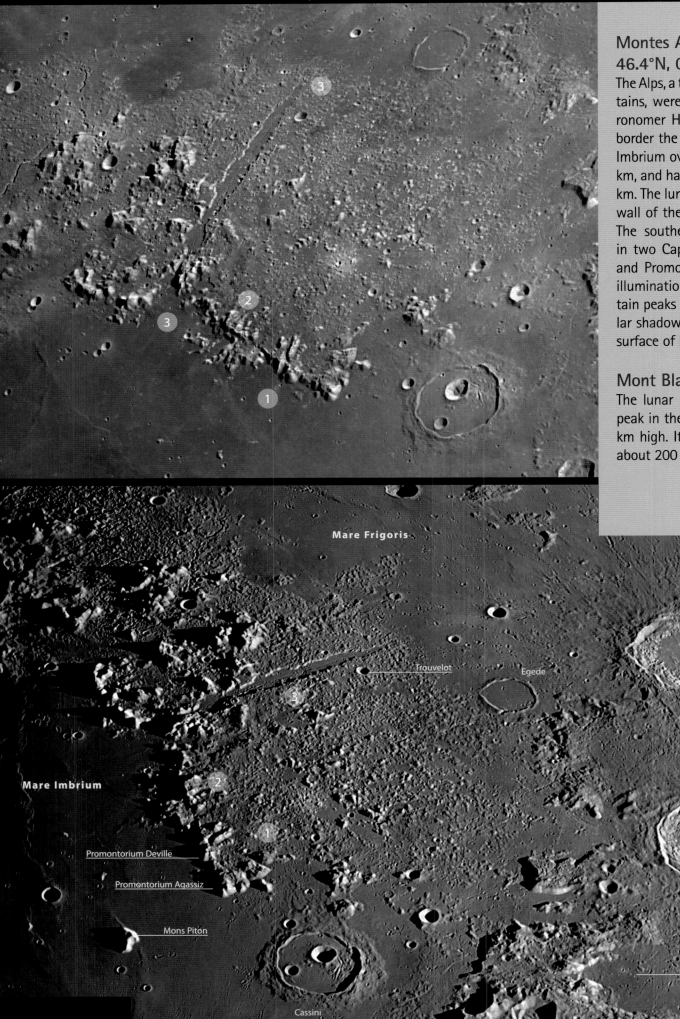

Montes Alpes
46.4°N, 0.8°W

The Alps, a tremendous range of mountains, were named by the Polish astronomer Hevelius (1611–1687). They border the northeastern area of Mare Imbrium over a distance of about 280 km, and have an average height of 2.4 km. The lunar Alps are a portion of the wall of the enormous Imbrium Basin. The southern end of the Alps ends in two Capes, Promontorium Agassiz and Promontorium Deville. With low illumination at sunrise, the mountain peaks throw numerous spectacular shadows onto to the lava-covered surface of Mare Imbrium.

Mont Blanc 45.4°N, 0.6°E

The lunar Mont Blanc is the highest peak in the Alpine chain at about 3.6 km high. Its surface area amounts to about 200 km².

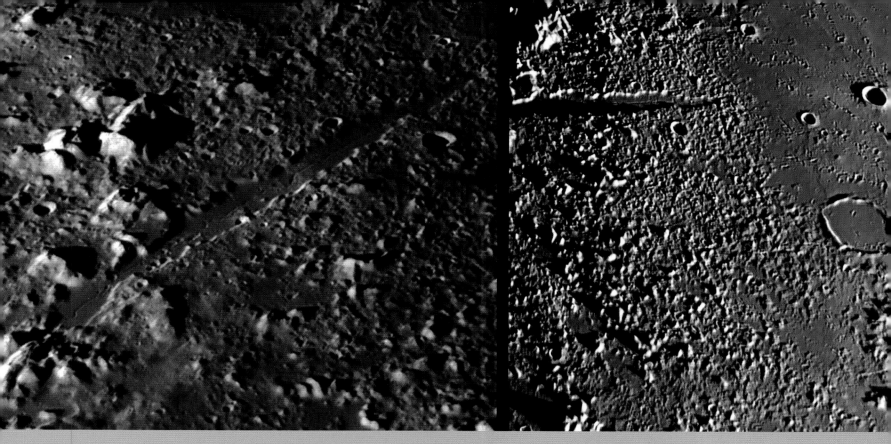

Vallis Alpes 48.5°N, 3.2°E ③

The Alpine Valley, with its overall length of just 170 km, is undoubtedly one of the most spectacular objects of lunar topography, and is a prominent landmark in the northern area of the Moon. It attains a maximum width of 11 km, and lies nearly radially to the centre of the Imbrium Basin, cutting the Alpine mountains into northwestern and southeastern sections. The Vallis Alpes is not a geological surface feature like valleys on Earth, which have been formed by running water and erosion. It consists of two, parallel tectonic fracture zones (similar to the Rift Valley in East Africa), between which the lunar surface has broken and subsided. It is actually a gigantic-sized, but typical, linear rille. The Alpine Valley probably arose at the same time as the Imbrium impact, and its

floor was slowly filled with lava flows that produced a level surface. In telescopes with apertures of about 20 cm or more – under good seeing conditions and appropriately favourable lighting – an extremely narrow rille, which is interrupted in a few places by crater pits, becomes visible. The most favourable observing conditions occur shortly before First or Last Quarter.

The central rille has a width of 0.55 km to 1 km as a maximum and the depth (determined from shadow measurements from Lunar Orbiter images) varies between 80 and 240 metres. It is probably a former lava tube with a collapsed roof, that ran down the rear slope of the Alps.

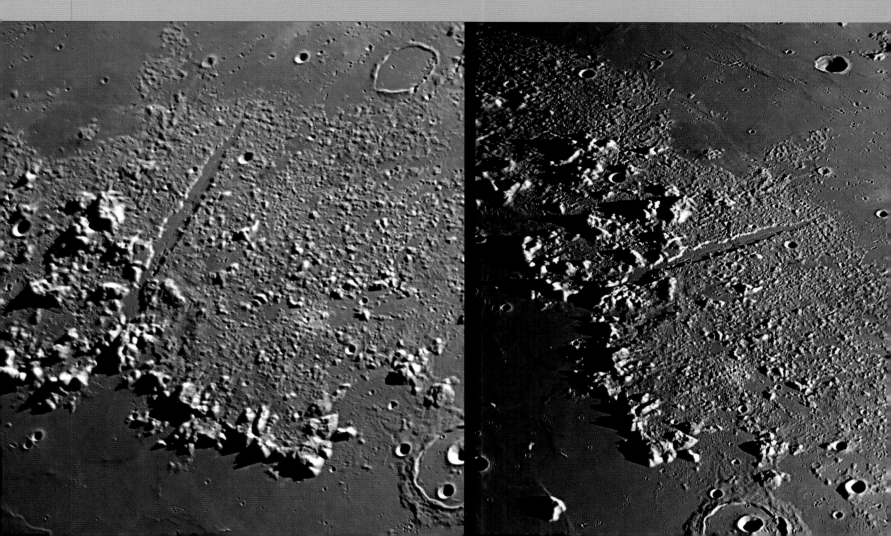

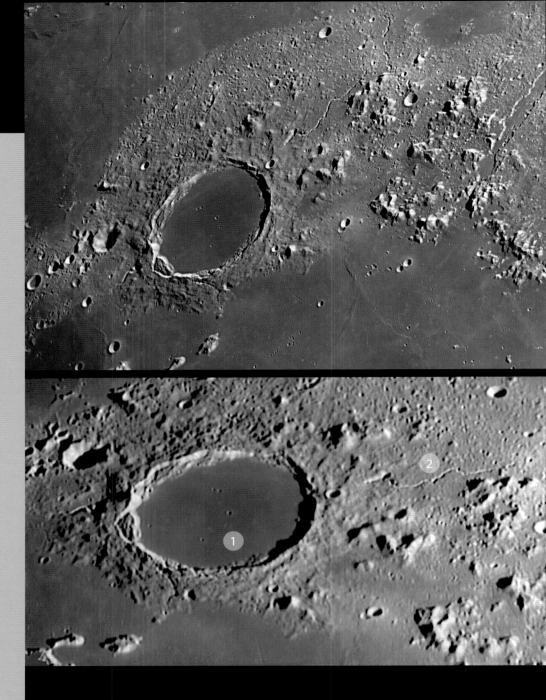

Plato 51.6°N, 9.4°W ➊

Plato is a large crater, 109 km in diameter, with an average depth of only 1 km. A few peaks on the eastern and western crater rim reach a height of 2 km. The crater is visible even with binoculars and is more-or-less divided in two at Half-Moon phases: half illuminated, and the other half lying in the darkness of the lunar night. Plato is sometimes referred to as 'the big black lake'. In fact, the crater floor of Plato is one of the darkest areas on the nearside of the Moon. Plato is nearly circular, but because of limb-foreshortening appears elliptical to an observer on Earth. Two large, triangular landslips are visible in the western edge of the crater. Similar wall or mountain landslides are also found in the craters Aristoteles and Gassendi.

Plato is extensively lava-flooded, and even the peaks of the central mountains – which must have been present in a crater of this size – lie below the lava surface. A normal crater of this size should be between 3 and 4 km deep. The upper surface of the lava appears com-

pletely flat, level, and without any features. Larger telescopes and good seeing conditions, however, enable five crater pits (about 2 km across) to be seen. Large telescopes with apertures of 300 mm and above and the use of webcam techniques show as many as 15 such pits. At sunrise (waxing Moon) the mountain peaks on the eastern ramparts throw long, sharply defined shadows on the floor of the crater, giving a highly spectacular sight. Over an observational period of one or two hours the changes in the lengths of the shadows may be clearly followed.

Directly adjacent to Plato on the south is a structure, which when illuminated near the terminator appears, at first sight, like a ghost crater. On old lunar maps this structure is named after the physicist Newton. In fact, it is not the remnants of a submerged or lava-flooded crater, but rather an optical illusion. As a result, the IAU has now accordingly given the name Newton to a crater near the South Pole. The structure is currently without a name, and is sometimes referred to as 'Ancient Newton'.

Rimae Plato 51.0°N, 2.0°W

A system of sinuous rilles, in three sections that are separated from one another. The main rille begins directly to the east of Plato, and is about 87 km long. It is probably a former lava tube is visible in medium-sized telescopes.

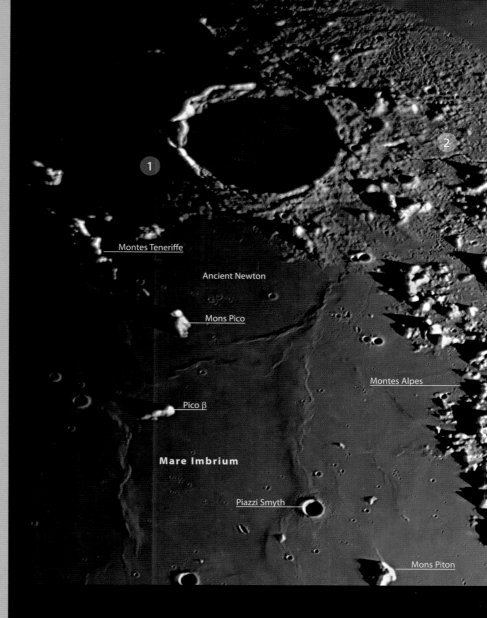

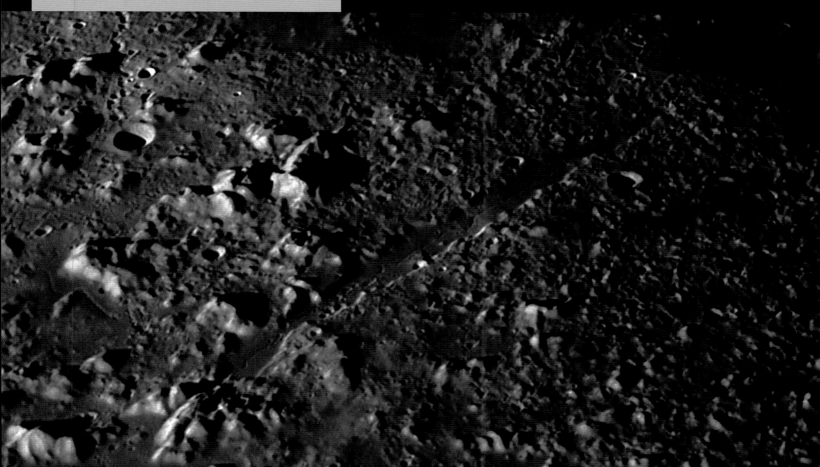

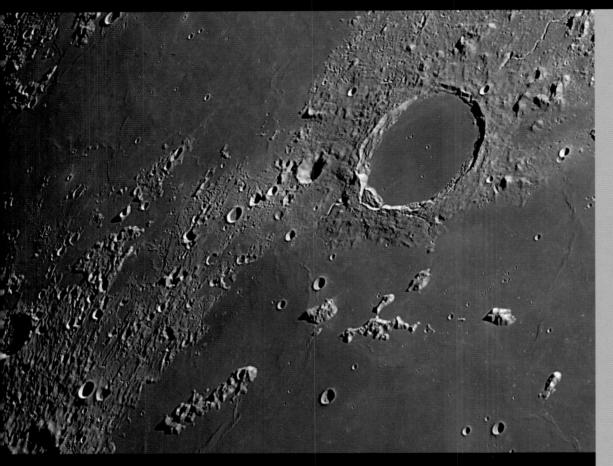

Montes Teneriffe ①
48.0°N, 13.0°W

Montes Teneriffe are a loose group of mountain peaks, which, with a height of about 2.4 km are all about the same height, and so do not differ from the height of Mons Pico. The mountains Montes Teneriffe, Montes Recti, and the individual mountains Mons Piton and Mons Pico are probably the towering remnants of the otherwise submerged inner wall of the Imbrium Basin.

Montes Recti ②
48.0°N, 20.0°W

A linear range of mountains, running directly east-west for about 90 km, and with a width of about 20 km. The highest peaks reach a height of 1.8 km above the level of the lava of Mare Imbrium. They are part of the inner wall of the Imbrium Basin.

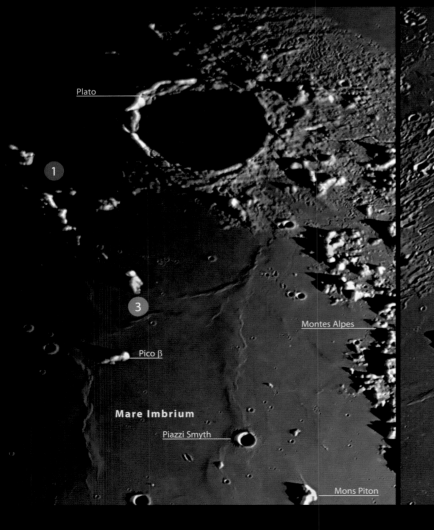

Plato

①

③

Montes Alpes

Pico β

Mare Imbrium

Piazzi Smyth

Mons Piton

Mare Frigoris

②

①

Mare Imbrium

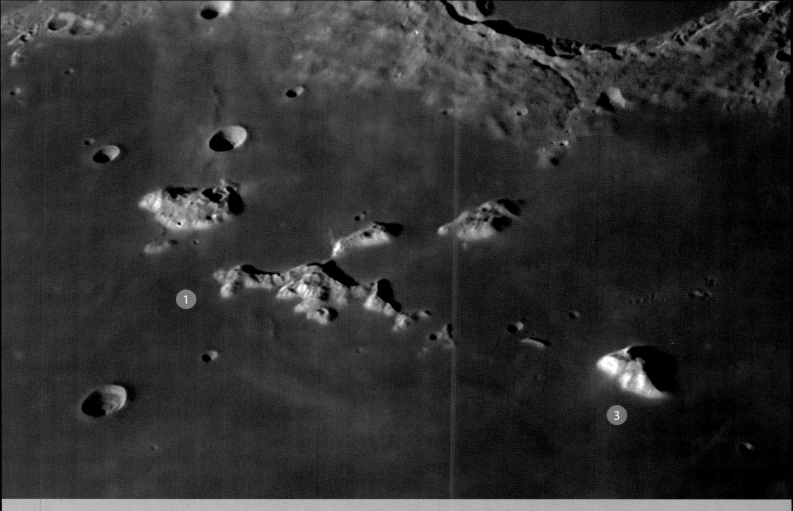

Mons Pico 46.0°N, 9.0°W

3

Mons Pico is an isolated mountain with an area at its base of 15 × 25 km and a height of 2.4 km. It lies exactly on the edge of 'Ancient Newton'. The mountain was named by Johann Hieronyus Schröter (1745–1816) after Pico del Teide on Teneriffe, a peak that he compared to the mountain on the Moon.

When Mons Pico is observed under grazing illumination, the mountain casts a spectacular, pitch-black shadow, as much as 90 km long, on the surrounding lava surface, and gives the appearance of an extremely steep, towering peak. Because the Moon has no atmosphere whatsoever, sunlight is not scattered and this intensifies the shadow. When the height is examined in relation to the base (15 × 25 km) it is found to be rather a considerable hill with relatively shallow slopes rather than a steep peak. Because of the small diameter of the Moon and the resulting steep curvature of the horizon, the mountain peak would disappear below the horizon at a distance of 90 km. The same applies to Mons Piton, which lies nearby, and which, with a height of 2.3 km, is almost as high as Mons Pico.

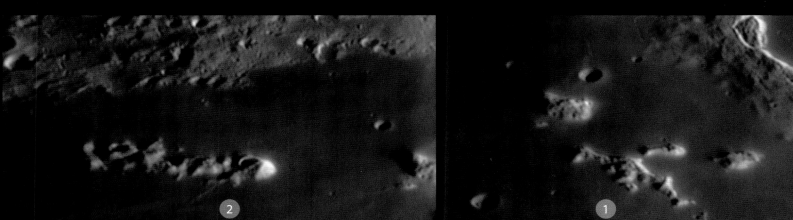

30 Archimedes

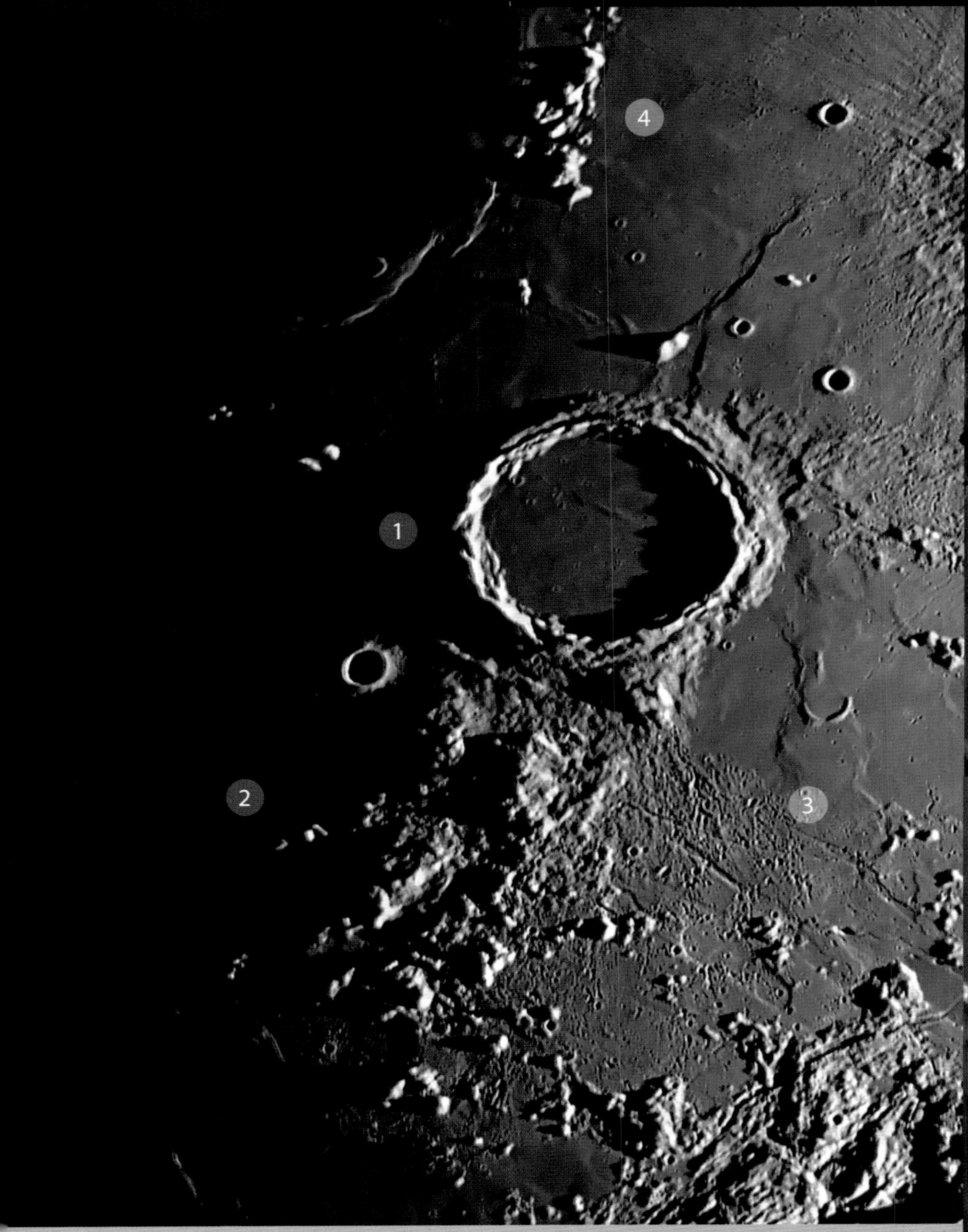

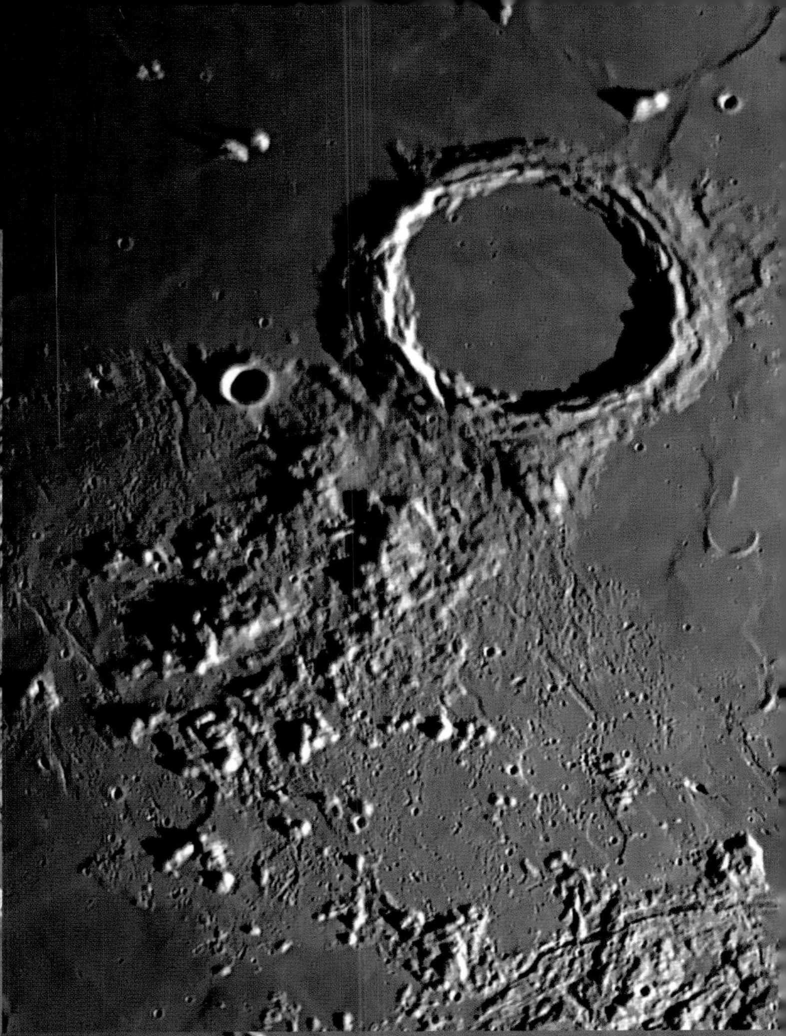

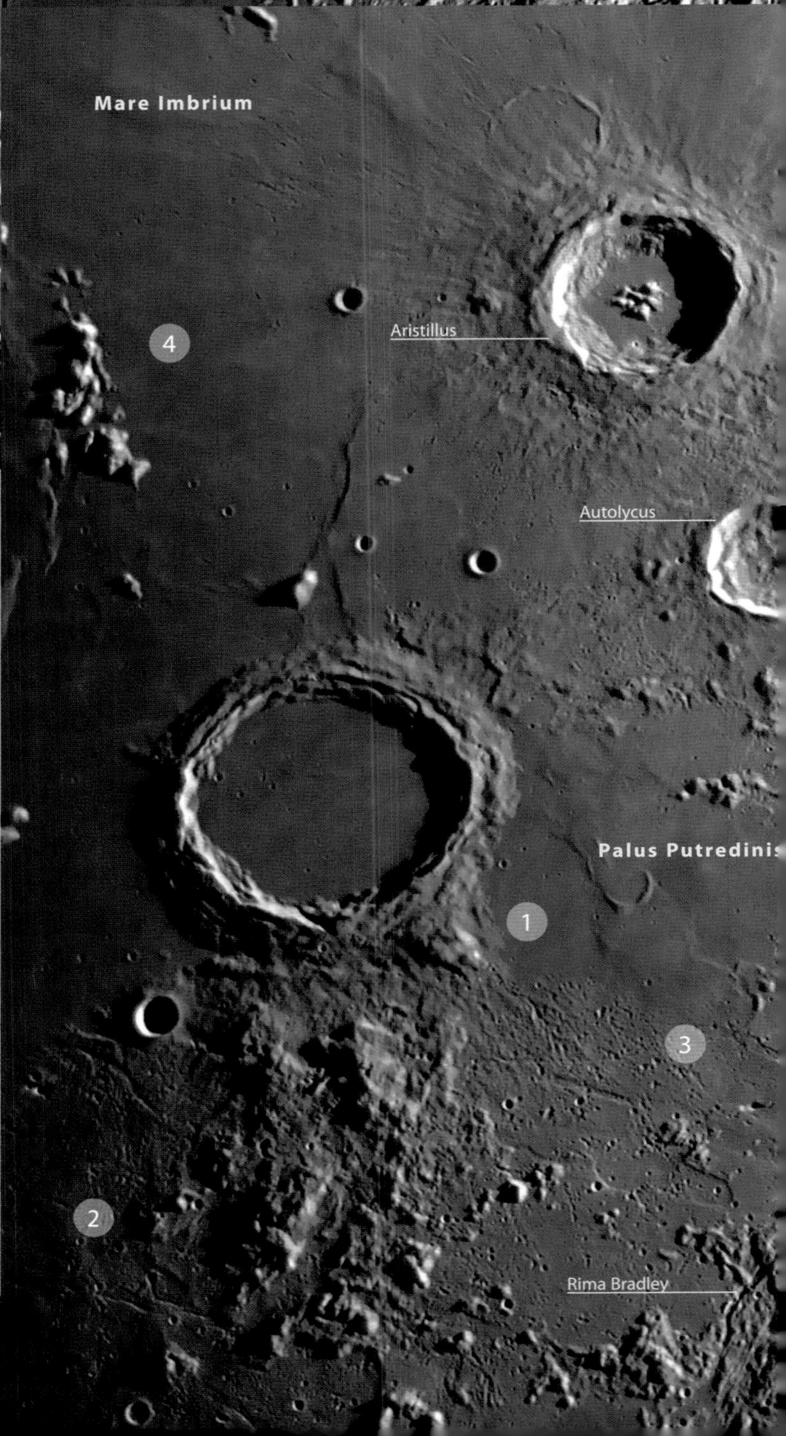

Mare Imbrium

Aristillus

Autolycus

Palus Putredinis

Rima Bradley

Archimedes 29.7°N, 4.0°W

Archimedes is one of the most conspicuous craters on the Mare Imbrium with a lava-flooded crater floor, and which was probably created during the early Imbrium period. The inner crater walls are stepped or terraced. The diameter of Archimedes is 82 km and the crater walls reach a height of 2.1 km above the floor. Its considerable diameter suggests that there are central peaks, but these have been completely submerged by the lava. The lava surface inside the crater appears smooth and completely level. Medium-sized instruments show, under favourable lighting conditions, two craterlets; Archimedes T, right over to the west, and Archimedes S, close to the eastern wall of the crater. Both have a diameter of just about 3 km. Large instruments show at least a dozen smaller crater pits. Stripes of bright ejecta, of unknown origin, lie across the lava surface. The ejecta melted by the impact also lies on the upper wall of the crater and gives it a thickened, rounded appearance. A prominent, triangular cape stretches out from the crater rim towards the southeast, to a distance of 30 km.

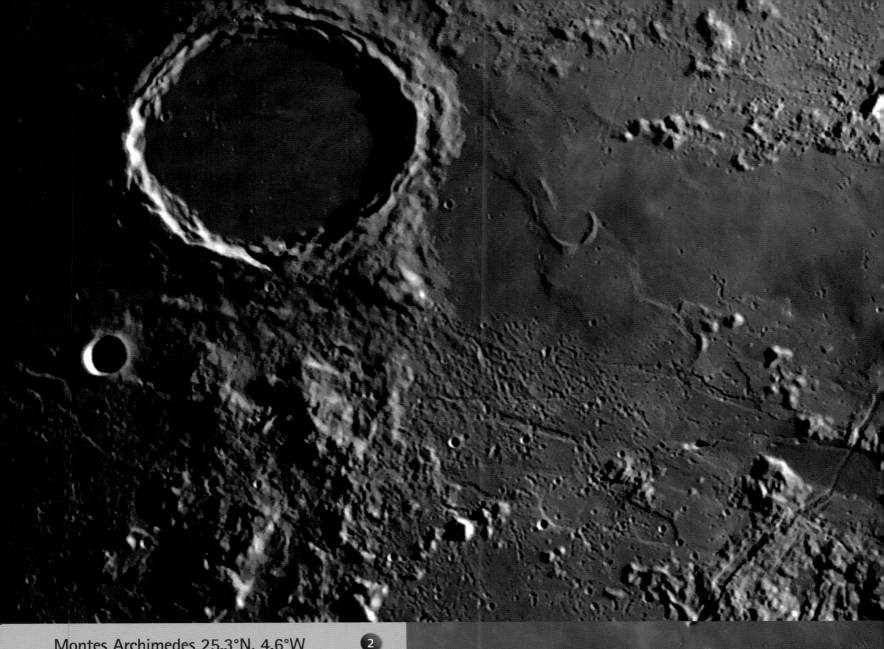

Montes Archimedes 25.3°N, 4.6°W ②

Montes Archimedes is a range of mountains that covers an area of about 160 × 140 km. The mountains directly adjoin the southern wall of Archimedes and, from there, run out in a southerly direction. They are a portion of the original inner wall of the Imbrium Basin, which had been heaped up to such a height that they were not submerged by the Imbrium lava.

Rimae Archimedes 26.6°N, 4.1°W ③

A system of narrow, linear rilles with an overall length of about 170 km, lying southeast of Archimedes. The begin at the triangular cape on the southern crater rim and stretch as far at the foothills of the Apennines. Two of the rilles are also visible in small telescopes, but all the others require large apertures.

Montes Spitzbergen 35.0°N, 5.0°W ④

A chain of mountains about 60 km long, lying north of Archimedes. Individual mountain peaks reach a height of 1.5 km above the surface of the lava. With low solar illumination (at sunrise or sunset) the mountain peaks cast spectacular, pitch-black, long shadows on the lava surface of Mare Imbrium. Directly to the west, under similar illumination, segments of a long mare wrinkle ridge are visible.

Montes Apenninus 18.9°N, 3.7°W ①

The Apennines are the greatest range of mountains on the Moon's nearside. The name was first used by the Polish astronomer and lunar observer Hevelius (1611–1687). Like the Alps, the Apennines are the remnants of the rim of the Imbrium Basin that have not been flooded by lava.

The Apennines stretch for a total length of about 600 km and individual mountain peaks reach heights of 5 km. The slopes towards the Mare Imbrium are relatively steep at inclinations of about 30°, whereas on the opposite side, towards the Mare Vaporum ('Sea of Vapours') and Sinus Aestuum ('Bay of Billows') they are significantly less. The Apennine hinterland probably consists largely of ejecta from the Imbrium impact. When illuminated at Full Moon, the Apennines appear very bright.

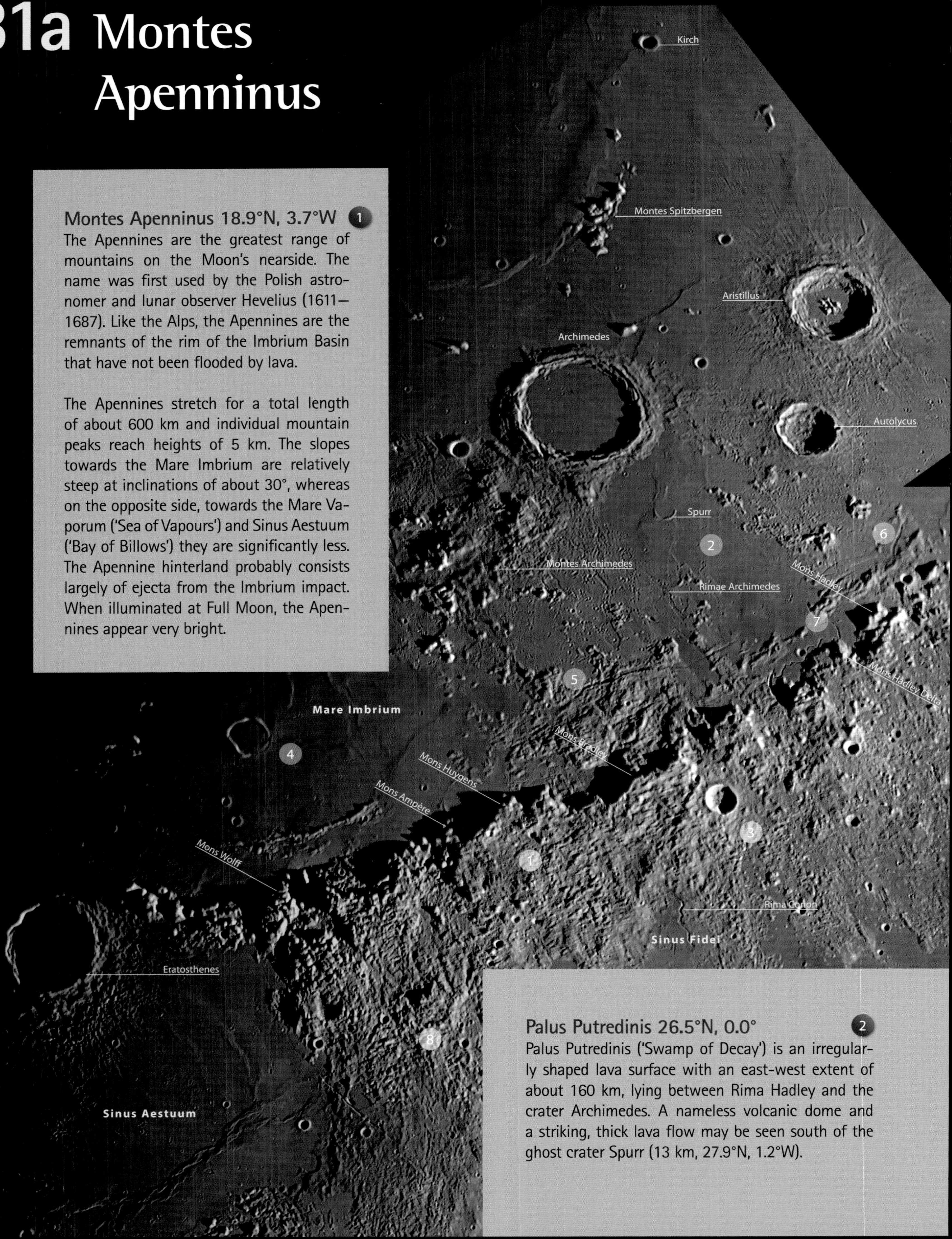

Kirch

Montes Spitzbergen

Aristillus

Archimedes

Autolycus

Spurr

②

⑥

Montes Archimedes

Mons Hadley

Rimae Archimedes

⑦

⑤

Mons Hadley Delta

Mare Imbrium

Mons Bradley

④

Mons Huygens

Mons Ampère

③

Mons Wolff

①

Rima Conon

Sinus Fidei

Eratosthenes

⑧

Palus Putredinis 26.5°N, 0.0° ②

Palus Putredinis ('Swamp of Decay') is an irregularly shaped lava surface with an east-west extent of about 160 km, lying between Rima Hadley and the crater Archimedes. A nameless volcanic dome and a striking, thick lava flow may be seen south of the ghost crater Spurr (13 km, 27.9°N, 1.2°W).

Sinus Aestuum

Conon 21.6°N, 2.0°O (3)

Conon is a very prominent crater, with a diameter of 21 km, with a sharp crater rim, lying behind the Apennines. It is possibly the result of a secondary impact caused by the main Imbrium impact. The broad rille, Rima Conon, snakes its way south of the crater in a zig-zag fashion for a distance of 45 km, and is visible in small instruments. It ends at Sinus Fidei (the 'Bay of Trust'). When illuminated by the Sun, the inner crater wall of Conon is highly reflective of sunlight, and appears extremely bright.

Wallace 20.3°N, 8.7°W (4)

Wallace is a crater, almost completely flooded by lava. Its floor appears completely smooth and level, without any detectable crater pits. The visible remainder of the crater wall appears almost square and slopes down below the Mare Imbrium lava on the southeastern side. When near the terminator, the wall of the crater casts a striking shadow, enabling its shape to be readily seen. Southeast of the crater and lying parallel to the foot of the Apennines, is an enlongated, nameless mountain ridge with an interesting structure.

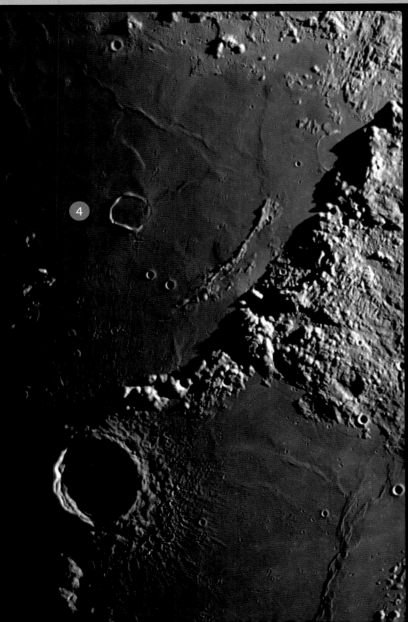

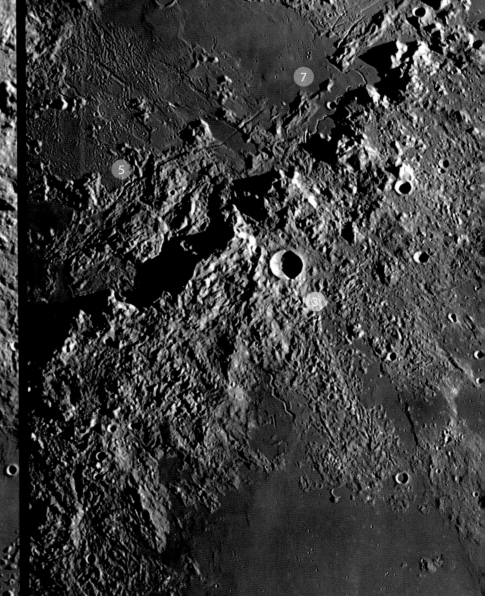

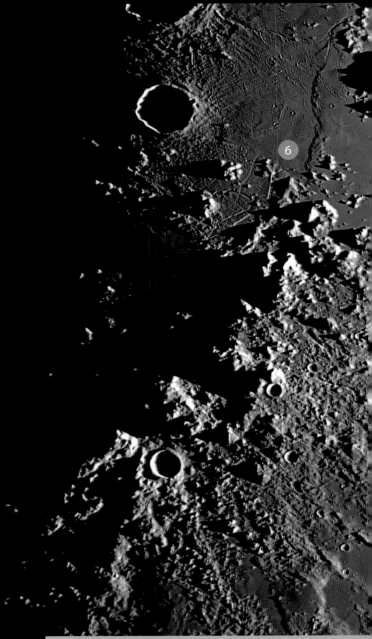

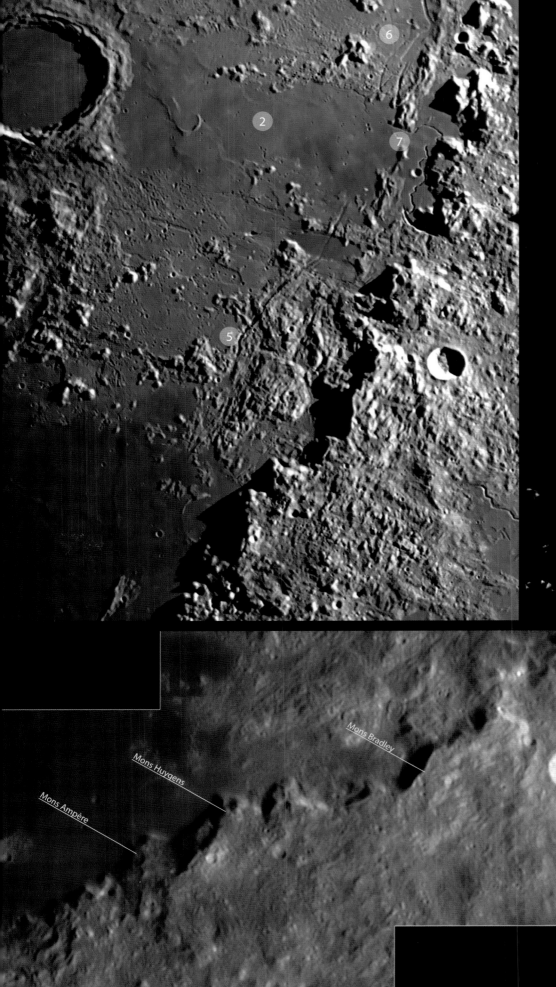

Mons Bradley

Mons Huygens

Mons Ampère

Rima Bradley 23.8°N, 1.2°W ⑤
A conspicuous, 3 km wide straight rille running parallel to the Apennines, 130 km long. It undoubtedly marks a fracture zone at the foot of the Apennines.

Rimae Fresnel 28.0°N, 4.0°E ⑥
A complex system of linear rilles with a total length of about 90 km, lying north-west of Mons Hadley. The rille system ends near Cape Fresnel (the northern outlier of the Apennines). Visually, it appears to be an extension of Hadley Rille, but geologically they are separate rille systems.

Apollo 15

Mons Hadley Delta

Bela

Rima Hadley 25.0°N, 3.0°E 7

Rima Hadley is a sinuous rille with an overall length of about 80 km. It begins at the crater Bela, which, with a size of 2 × 11 km, under particular illumination, often appears in a telescope more like a sickle-shaped mountain ridge than a crater. The rille follows the foot of the Apennines, until it turns through 90° and ends at the foot of the mountain Mons Hadley Delta. The rille is a typical lava channel with a width of between 1 and 2 km and a depth of about 300 m. On 30 July 1971, Apollo 15 landed near the rille. The rille was thoroughly investigated, photographed and measured. A descent into the rille by the crew of Apollo 15 was forbidden by the flight directors because of the steep gradients involved.

Marco Polo 15.4°N, 2.0°W 8

The remnants of an elongated crater, measuring 21 × 28 km, lying behind the Apennine Front.

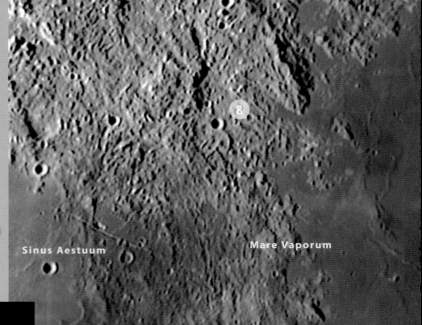

Sinus Aestuum

Mare Vaporum

Aristillus

Autolycus

Archimedes

Mare Serenitatis

Timocharis

Sulpicius-Gallus-Formation

Conon

5

3

Mare Imbrium

2

Montes Apenninus

Rima Conon

4

1

Eratosthenes

Boscovich

Julius Caesar

Sinus Aestuum

Rima Hygihus

Rima Ariadaeus

Ukert

Sinus Medii

Agrippa

Mare Vaporum 13.0°N, 3.5°E ⓵

Mare Vaporum ('Sea of Vapours') has an irregular, rounded shape with a diameter of 240 km and a lava-covered surface that amounts to about 55 000 km². Mare Vaporum lies to the southeast of the Apennine mountains. In the south, Mare Vaporum merges, without any visible boundary, into the lava plain that is Sinus Medii. The southwestern and southeastern boundaries betwen Mare Vaporum and the highland regions are marked by secondary impacts from the great Imbrium impact (forming part of the Imbrium Sculpture). To the northwest lie Sinus Fidei and Lacus Felicitatis on the borders with the Apennine mountains.

Lacus Felicitatis 19.0°N, 5.0°E ⓶

Lacus Felicitatis ('Lake of Happiness') is a smaller, irregularly shaped lava area of only about 90 km in extent. It lies southeast of the crater Conon and east of Rima Conon. Scientists believe that they have found evidence here that part of this area still exhibits volcanic activity in the form of outgassing. Proof came in 2006 from a small area, named Ina, that lies on the southeastern edge of Lacus Felicitatis. Ina is a semicircular depression (shaped somewhat like the heel of a shoe), whose coordinates are 18.6°N, 5.3°E, and with a diameter of only about 3 km (about 1.5 arcseconds) and a depth of about 30 m. Even in large amateur telescopes the area is only partially visible.

Menelaus
16.3°N, 16.0°E

Menelaus is another young, very prominent crater, 26 km in diameter, with a sharp crater rim and central peaks. Like Manilius it is the origin of a bright ray system. One of the rays is more than 1000 km long and crosses right across Mare Serenitatis and in doing so also crosses the crater Bessel. Other rays are visible both to the southwest and southeast, between Menelaus and the craters Auwers (20 km) and Daubrée (14 km).

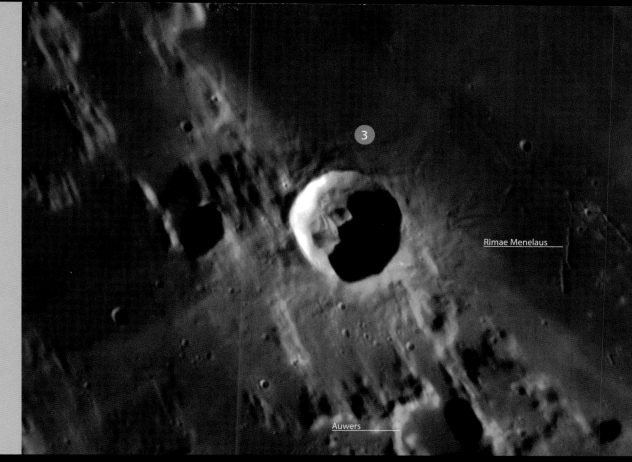

Rimae Menelaus

Auwers

Manilius
14.5°N, 9.1°E

A conspicuous young crater with a sharp crater wall, terraced inner walls and central peaks. It has a diameter of 38 km and is the source of bright rays of ejecta that become clearly visible at solar elevations of 35° or greater. One of the rays stretches for 400 km. Beginning at the western wall of the crater, the ejecta covers a narrow patch of Mare Vaporum, crosses the area of Dark Mantle Deposit in the highlands between Mare Vaporum and Sinus Aestuum, and end north of the crater Bode in Sinus Aestuum.

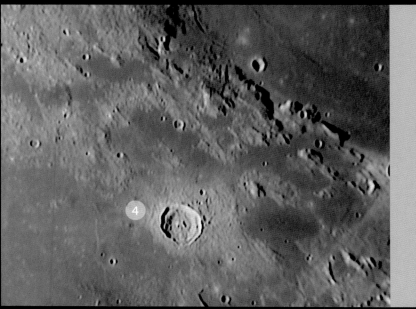

Montes Haemus 19.9°N, 9.2°E

A chain of mountains, about 400 km long, bordering Mare Serenitatis. The range is a remnant of the wall of the Serenitatis Basin. In the vicinity of the mountains, to the southwest, there are several small, irregularly shaped areas of lava: Lacus Odii ('Lake of Hatred'), Lacus Doloris ('Lake of Sorrow'), Lacus Gaudii ('Lake of Joy'), Lacus Hiemalis ('Lake of Winter'), Lacus Lenitatis ('Lake of Softness') and Lacus Felicitatis ('Lake of Happiness').

Menelaus

Manilius

Boscovich

Sosigenes

Julius Caesar

1

Rimae Sosigenes

Silberschlag

3

1

Agrippa H

Ariadaeus

Mare Tranquillitatis

5

4

2

Dionysius

Rima Ariadaeus 6.4°N, 14.0°E   ①

Together with the Alpine Valley, Rima Ariadaeus is one of the most impressive examples of a linear rille. Based on its width of up to 7 km it is almost always described as a valley or a graben. Rima Ariadaeus is easy to observe with small telescopes.

Rima Ariadaeus is a depression between two parallel fracture zones or faults. The rille runs with a nearly east-west orientation, almost radially to the direction of the centre of the Imbrium Basin. It is about 250 km long and with a depth of 500 m, not particularly deep. After about one third of its course (from the western end) it is interrupted by a mountain ridge (lying almost at a right angle to the rille), which begins to the north of the rille and continues south as far as the crater Tempel. South of the rille and east of the ridge lies the small, circular crater Silberschlag (13 km, 6.2°N, 12.5°E). The name of the rille is derived from the crater Ariadaeus, which lies at the eastern end of Rima Ariadaeus. Here the rille bends slightly towards the south, crosses into the lave of Mare Tranquillitatis and ends at the Rimae Ritter rille system. Large telescopes using webcam techniques reveal a few crater pits on the floor of the rille.

Godin 1.8°N, 10.2°E
Agrippa 4.1°N, 10.5°E
Tempel 3.9°N, 11.9°E

Godin is a prominent crater of about 34 km in diameter, and somewhat irregular in shape. Near Full-Moon phase it shows a ray system. The crater's floor is roughly fissured and is crossed by a range of mountains with a central peak. When near the terminator, the walls of the crater that are illuminated are highly reflective.

Agrippa lies directly north of Godin. It is a 44-km crater of the Eratosthenian period on the lunar timescale. It has a V-shaped central peak, with the point of the V towards the north. The northern crater wall has been breached by the small crater Agrippa H (6 km). As with Godin, those walls of the crater that are illuminated by grazing light appear very bright.

Tempel is an almost completely destroyed crater, 45 km in diameter, directly east of Agrippa. The pair of craters Agrippa and Tempel present a good example of the differing state between an old (Tempel) and a young crater (Agrippa).

Cayley
4.0°N, 15.1°E

A circular crater of only 14 km diameter and with what is, for this diameter, an unusual depth of 3 km. The ratio of depth to diameter here is 1:5, whereas the normal ratio for this diameter of crater is between 1:12 and 1:24.

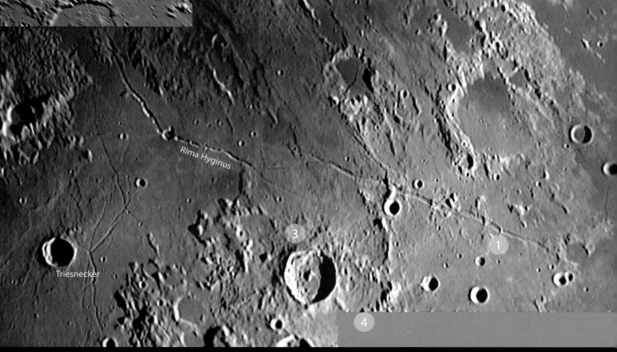

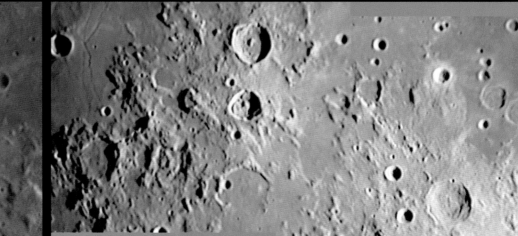

Mare Vaporum

Boscovich P

Boscovich

Triesnecker

Sinus Medii

Rima Hyginus 7.4°N, 7.8°E

Rima Hyginus is an extraordinary feature and, together with Rimae Triesnecker, is the best-known of all rilles. It consists of two linear sections, each of about 11 km in length, and is interrupted in the centre by the crater Hyginus (11 km, 7.8°N, 6.3°E). Hyginus is extended on the north by an unnamed small crater of about 4 km in diameter.

Both sections of the rille are about 400 m deep. The eastern section runs nearly parallel to Rima Ariadaeus, whereas the western section bends at the crater Hyginus by about 35° towards the north. This section is interrupted by nine craterlets, which lie directly in the rille. Rima Hyginus is probably an old lava channel and the craterlets may have been responsible for repeated eruptions of volcanic ash. This suspicion is strengthened by the fact that there are many nearby areas that are coated with dark volcanic ash (they are DMD regions). The crater Hyginus itself also appears to be of volcanic origin, because there is no form of crater wall and the floor is absolutely flat and level – it appears as if it were a cylinder punched out of the lunar surface.

About 35 km north of the crater Hyginus lies a small mountain, that appears spiral, rather like the letter 'e'. The mountain has the former official name of 'Schneckenberg' ('Snail Mountain', at 9.0°N, 6.5°E). East of this lies the extensively eroded crater Boscovich (46 km, 9.8°N, 11.1°E). On its floor is the rille system Rimae Boscovich (see image Section 33), which consists of two sections that lie at right angles to one another. Adjacent to the northwest is Boscovich P, a nearly perfectly rectangular, lava-flooded feature covering an area approximately 30 × 80 km.

Schneckenberg

Hyginus

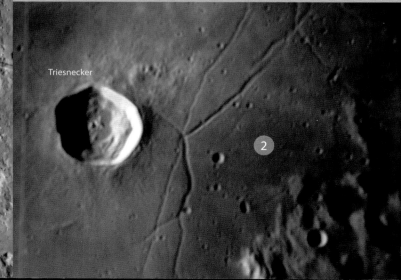

Rimae Triesnecker 4.3°N, 4.6°E

A greatly branching and very complex system of rilles, many of which are long and thin. The system lies almost at right angles to Rima Ariadaeus and Rima Hyginus. The rilles stretch in a north-south direction over a length of almost 200 km. These are former lava channels, which probably transported the lava that formed Sinus Medii. The main sections of Rimae Triesnecker are actually visible in small telescopes. The name derives from the complex crater Triesnecker (4.2°N, 3.6°E), with a diameter of 26 km and a poorly developed ray system.

The whole region between Rima Hyginus in the west and the edge of Mare Tranquillitatis shows broad surface structures, oriented radially to the Imbrium Basin. They are part of the Imbrium Sculpture.

Triesnecker

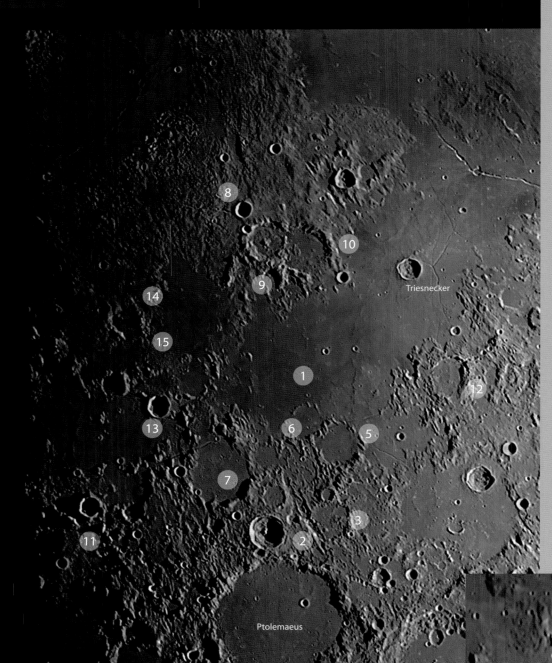

Triesnecker

Ptolemaeus

Sinus Medii 0.0°, 0.0° ①

Sinus Medii ('Central Bay'), is a moderately large lava area about 350 km across. It lies nearly in the centre of the nearside of the Moon and thus contains the origin of the system of lunar coordinates. There are many interesting surface features in this area – which are even visible in small telescopes.

Herschel 5.7°S, 2.1°W ②

Herschel is a prominent crater, 40 km in diameter, with a rough crater floor, eccentric small central peak, and terraced inner crater walls. It lies directly north of the large complex crater Ptolomaeus.

Gyldén 5.3°S, 0.3°W ③

An almost completely destroyed crater, 47 km in diameter. The western wall is breached by a wide valley, caused by secondary impacts and shock waves from the Imbrium impact.

Bruce 1.1°N, 0.4°E ④

A circular crater, 6 km in diameter, lying closest to the origin of the lunar coordinate system.

Mösting A

Ptolemaeus

Réaumur 2.4°S, 0.7°E ⑤

An almost completely destroyed crater, 52 km in diameter, with partially submerged crater walls in the north. Rima Réaumur (about 30 km) lies to the east and Rima Oppolzer (about 95 km) to the north of the crater remnant. Both rilles are surface fractures, that is, linear rilles. Adjoining them on the west is the 80-km long Rimae Flammarion system.

Oppolzer 1.5°S, 0.5°W ⑥

A crater remnant, lying northwest of Réaumur, and separated from it by Rima Oppolzer. Only rudimentary portions of Oppolzer's northern crater wall still exist.

Flammarion 3.4°S, 3.7°W ⑦

A crater, 74 km in diameter. The western wall is broken by the crater Mösting A (13 km). The northern crater wall is completely submerged by lava and is crossed by the Rima Flammarion rille (80 km).

A small crater, 18 km in diameter. To the northeast lies the Rimae Bode rille system, a sinuous rille (70 km, 10.0°N, 4.0°E), which is visible only in large instruments.

Pallas 5.5°N, 1.6°W (9)
Murchison 5.1°N, 0.1°W (10)

Pallas and Murchison are two craters, similar in size, on the northwestern edge of Sinus Medii, with diameters of 46 and 57 km, respectively. They are touching one another (oriented west-east) and the crater walls partially overlap. With some imagination the pair may be envisaged as looking like the Greek letter θ. Both craters have greatly destroyed crater walls and Pallas has a central peak. The crater wall of Murchison is breached in the southeast and the floor of the crater is open to Sinus Medii. Towards the interior of the crater there is a large, nameless lunar dome with a summit crater (a caldera), the shape of which closely resembles that of an Icelandic shield volcano.

Lalande 4.4°S, 8.6°W (11)

A prominent crater, 24 km in diameter, and with slightly angular walls. Under Full-Moon illumination conditions it is the centre of a ray system.

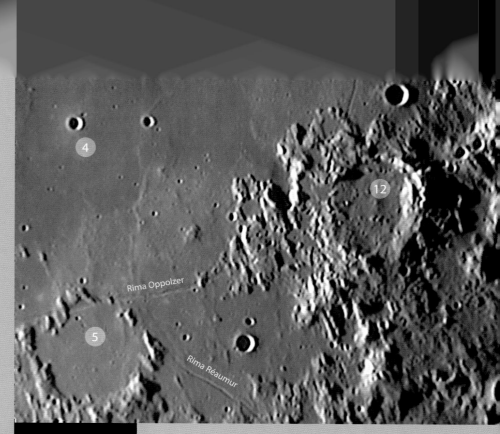

Rima Oppolzer

Rima Réaumur

Rhaeticus 0.0°, 4.9°E (12)

Rhaeticus, 43 × 49 km in size, is a large, irregularly shaped crater, which has been largely destroyed. Running north-south across the crater floor is a low ridge with the indications of a central peak. The centre of the crater lies almost exactly on the lunar equator. About 200 km to the west are the landing sites of Surveyor 6 and Surveyor 4. Surveyor 4 was a failure, the planned soft landing failed, and the probe smashed into the lunar surface.

Mösting 0.7°S, 5.9°W (13)

Like Lalande, Mösting is also a prominent crater with a diameter of 24 km and an angular crater wall. The crater's floor is rough and the inner walls are terraced.

Mösting A is a circular, bright crater, 13 km in diameter. It lies on the western wall of Flammarion and is a basic reference point for the system of selenographic coordinates (after Davies, 1987). Its coordinates are 3°12'43.2"S and 5°12'39.6"W.

Schröter 2.6°N, 7.0°W (14)
Rima Schröter 1.0°N, 6.0°W (15)

Schröter is a conspicuous, lava filled crater, 35 km in diameter. The southern crater wall has been buried by lava. To its south lies Rima Schröter, a linear rille, difficult to observe, which stretches for 40 km in a north-south direction. It begins at a crater pit and ends between two mountains (similar to Rima Birt). Despite its linear nature it may be a lava channel.

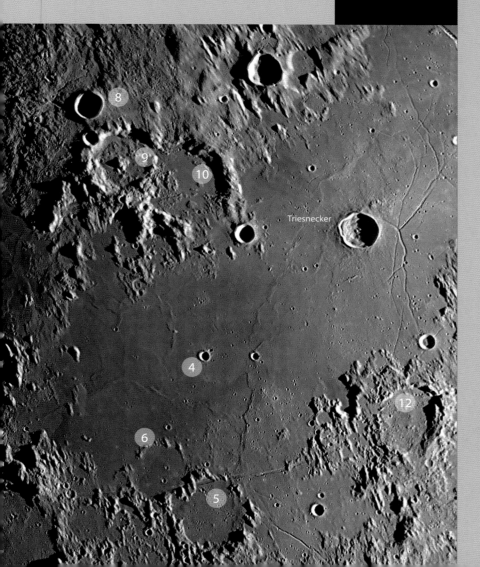

Triesnecker

Hipparchus 5.1°S, 5.2°E

Hipparchus is a giant, heavily ruined crater, with a diameter of 138 km and a depth of 3 km, which arose in the Pre-Nectarian period. The southeast crater wall is broken split by two deep, broad gaps. The northwestern wall of the crater is likewise largely destroyed, and exhibits wide gorges that are open towards Rima Réaumur (about 30 km long) and to Sinus Medii. Rima Réaumur encounters Rima Oppolzer at almost a right angle. Both are linear fracture zones, probably associated with the lava surface of Sinus Medii.

The floor of Hipparchus exhibits mountain peaks, small hills, crater pits, ghost craters and the 30-km, well-preserved, young crater Horrocks. The southwestern crater wall of Hipparchus is almost completely destroyed, and is covered with a group of both large and small craters.

The crater walls of Hipparchus generally consist of a series of holes, furrows, rilles, hills and ridges. The pattern that these features exhibit is radial to the centre of the Imbrium Basin. They are part of the Imbrium Sculpture and may be followed over a distance of more than 1000 km.

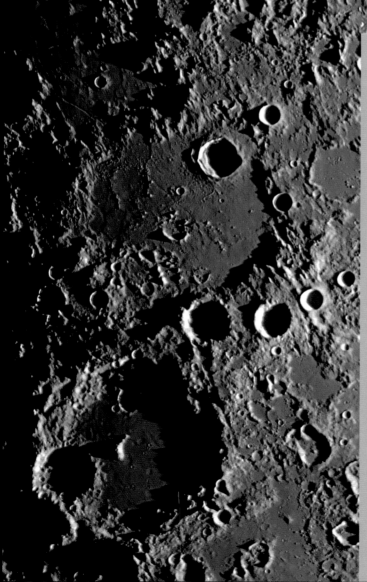

Albategnius 11.7°S, 4.3°E

Albategnius is a prominent, very conspicuous complex crater with a diameter of 114 km, an imposing central peak and a flat, smooth crater floor. The inner walls have been heavily eroded by landslides and intervening valleys, and is saturated with craterlets. The western crater wall is overlapped by the 44-km wide crater Klein. Under grazing illumination at sunrise (waxing Moon), numerous, very shallow, circular depressions are visible. Two of these lie directly to the east of the central peak, and two more to the south. The highest summit on the central peak has a pit that is about 1 km in diameter.

Hipparchus and Albategnius form a pair of craters with many interesting features, and are an observational highlight, regardless of telescope aperture.

Horrocks 4.0°S, 5.9°E
Pickering 2.9°S, 7.0°E

Horrocks, 30 km in diameter, is a young, extremely well-preserved impact crater, lying on the northwestern edge of Hipparchus. Pickering is a circular crater, 15 km in diameter, named after E.C. Pickering, the first Director of Harvard College Observatory.

Halley 8.0°S, 5.7°E ⑦
Müller 7.6°S, 2.1°E ⑧

Halley is a prominent young crater with a diameter of 36 km. A chain of very noticeable smaller craters runs in a northwesterly direction from the crater Müller (20 km) towards Ptolomaeus. The crater chain begins with Müller A (on the southern crater wall) and ends with the crater Ptolomaeus G. Under high solar illumination, the small crater Hind C (directly south of Hind, 7.9°S, 7.4°E) shows a readily visible bright ray system.

Lade 1.3°S, 10.1°E ⑤

Lade is a crater that is slightly pentagonal in shape, with a diameter of 55 km. The southern crater wall has been buried beneath lava, as has the whole floor of the crater. On the inner northwestern wall is the crater Lade M (12 km), which is also filled with lava right to the edge. The lunar coordinate system's equator lies between Lade and the crater Godin (35 km, 1.8°N, 10.2°E).

Vogel 15.1°S, 5.9°E ⑥

A crater 26 km across, and with the smaller craters Vogel A, B and C. Lying almost directly south of Albategnius, the crater forms an interesting trio of craters with Argelander (34 km, 16.5°S, 5.8°E) and Airy (37 km, 18.1°S, 5.7°E) with a sort of bottle- or gourd-like shape. The trio possibly arose from secondary impacts, created during the formation of the Imbrium Basin.

Ptolemaeus 9.3°S, 1.9°W ①

Ptolomaeus is a highly distinctive, very old large crater with a diameter of 164 km. The crater wall is slightly polygonal, and because of its extreme age is heavily eroded. It rises about 2.4 km above the floor of the crater. As with many other structures of this size, a gravitational anomaly (a mascon) is measured over Ptolomaeus. The crater's floor appears flat and smooth in small telescopes, but in larger instruments a wealth of small craters and pits becomes visible. On the northeastern floor there is the prominent crater Ammonius, with a diameter of 8 km, and directly adjoining it to the north is Ptolomaeus B, a shallow, saucer-shaped depression and, even farther north, an additional, unnamed depression.

The eastern crater wall has a peak, which casts a conspicuous shadow directly across Ammonius at sunrise.

On the western part of the crater's floor and under an extremely low angle of illumination, two very shallow, circular depressions may be observed. Directly outside the northeastern wall of the crater there is a short, but very prominent, nameless chain of craters. The appearance of the floor of Ptolomaeus changes dramatically over the course of a lunation. At First and Last Quarters the crater floor appears relatively dark, but near Full Moon becomes very bright.

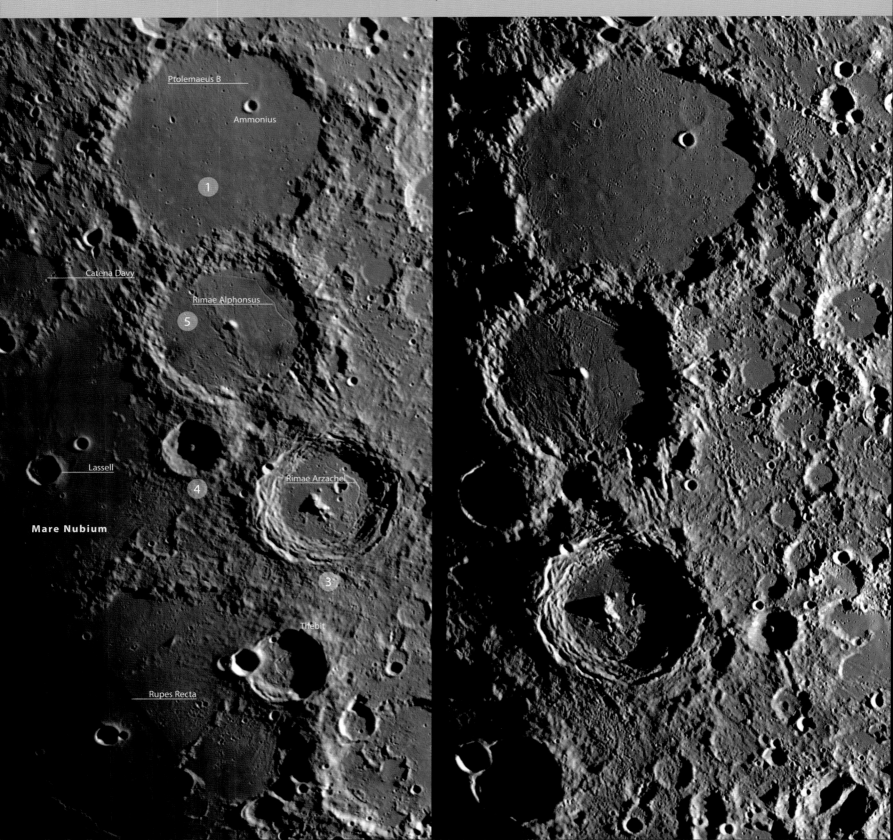

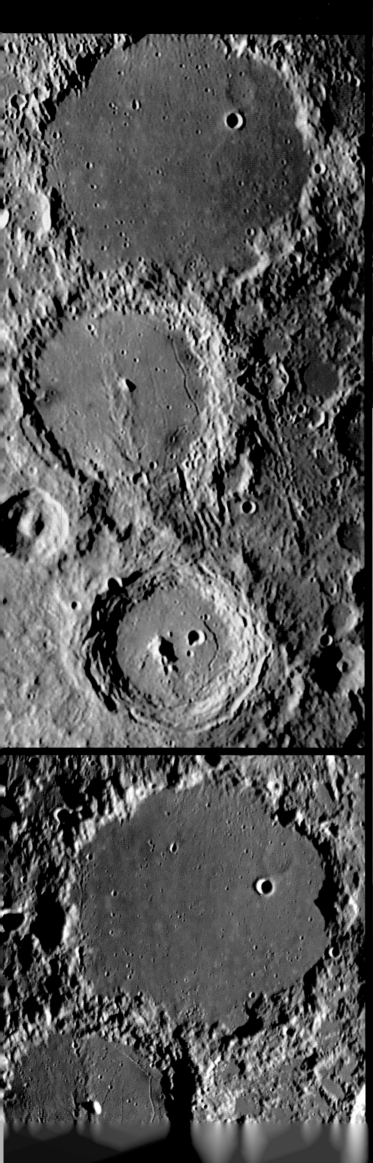

Davy 11.8°S, 8.1°W

②

Davy is a small, but conspicuous crater, 31 km in diameter. It lies west of Ptolomaeus, on the northeastern outlier of Mare Nubium. Davy A (15 km) breaches Davy's southeastern crater wall.

Directly adjoining it on the north lies the significantly larger crater Davy Y (about 60 km, 11.0°S, 7.0°W), on the floor of which is the well-known crater chain Catena Davy, a series of craterlets, which stretches for a distance of about 50 km.

Catena Davy consists of about 20, non-overlapping individual craters with diameters between 1 and 3 kilometres. Because of the small crater diameters, observation requires the use of a large telescope. Such chains of craters probably arose when an asteroid-like body with a relatively loose consistency (perhaps similar to a cometary nucleus), with a trajectory at a low angle to the lunar surface, was broken into fragments through gravitational forces shortly before the impact, or was previously a swarm of smaller bodies with closely similar flight paths.

The largest craters in the chain are Davy C (3.4 km diameter, 550 m deep) at the western end of the chain, and Davy YA, roughly in the centre.

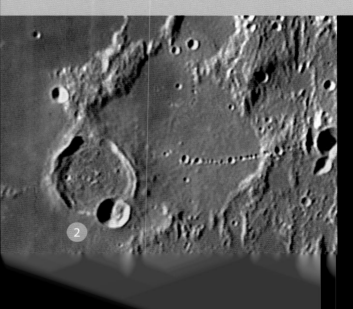

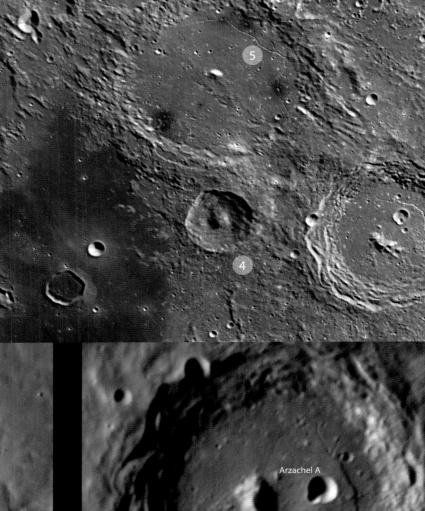

Arzachel 18.2°S, 1.9°W   3

Arzachel is younger than Ptolomaeus and Alphonsus. It has a diameter of 96 km. When
near the terminator it is an observational
highlight for any size of instrument. The inner crater walls are markedly terraced, with a
sharp-edged structure, and the long, slightly
eccentrically placed central peaks reach 1.5
km above the crater's floor. Many craterlets
and the 10-km crater Arzachel A lie on the
floor. Arzachel also has a rille system, Rimae
Arzachel. The main rille has a length of about
50 km, is significantly wider than the rilles in
Alphonsus and follows the curvature of the
eastern inner crater wall.

Alpetragius 16.0°S, 4.5°W   4

Alpetragius is a remarkable looking, circular crater, 39 km in diameter. For its size, it
is unusually deep. Its central peak is conical
in shape, and is conspicuously large when
compared with the crater's diameter. In all
probability, the central peak created by the
impact was subsequently modified and en-
larged through volcanism. Many craters in
the vicinity show these broad, conical and
unusually large central peaks.

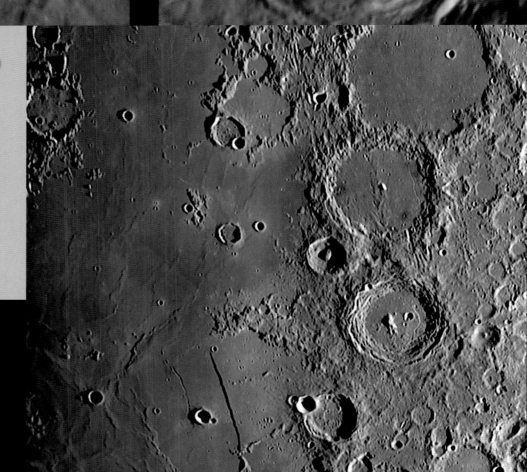

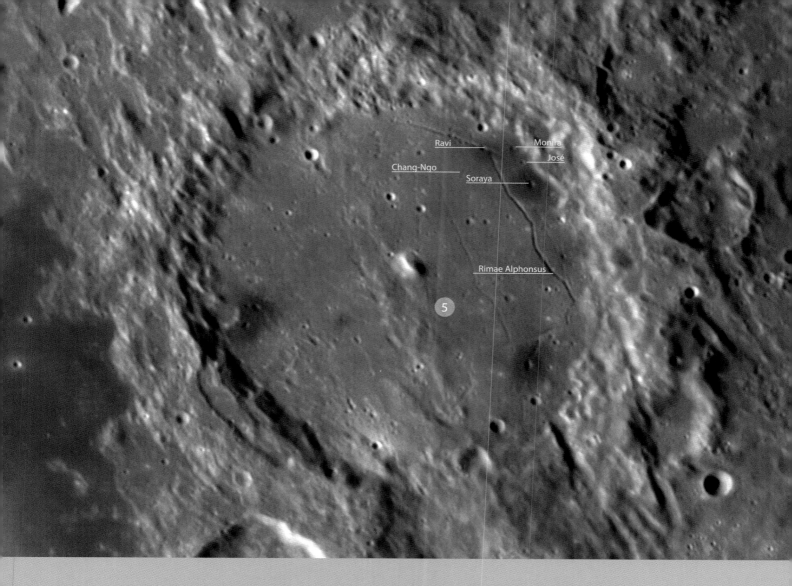

Labels on image: Ravi, Monira, José, Chang-Ngo, Soraya, Rimae Alphonsus, 5

Alphonsus 13.4°S, 2.8°W ⑤

Alphonsus is undoubtedly one of the most interesting craters on the nearside of the Moon. On old Moon maps it was described as a ring mountain. The prominent peak of the central mountain is visible, slightly north of the centre of the crater. A distinctive rille system, Rimae Alphonsus, lies on the eastern crater floor. Along the inner crater wall (both on the west and on the east) some very interesting crater pits may be observed. These are surrounded by a dark halo (i.e., are Dark Halo Craters). To a certain extent they are related to the rilles, from which it may be assumed that the rilles are former lava channels. (In the southeast, one rille branches at just such a crater pit.) A low ridge, similar to a mare wrinkle ridge crosses the whole crater floor in a north-south direction, and divides it into eastern and western sides. Alphonsus belong to the class of FFC craters. Alphonsus was formed in the Nectarian period, and was repeatedly mentioned in connection with LTP events. In March 1965 the Moon probe Ranger 7 crash-landed on the crater floor north of the central peak, sending television pictures back to Earth until the impact.

At least 10 crater pits, surrounded by dark haloes of volcanic ash, may be observed on the floor of Alphonsus with large telescope apertures (> 250 mm), including:

Ravi (2.5 km), 12.5°S, 1.9°W
Monira (2 km), 12.6°S, 1.7°W
José (2 km), 12.7°S, 1.6°W
Soraya (2 km), 12.9°S, 1.6°W.

⑤ These craters are undoubtedly of volcanic origin. The dark haloes consist of dark vitreous ash, produced by eruptive pyroclastic volcanism with lava having a high gas content. The adjoining rilles were originally faults, but which were transformed into lava channels by eruptive volcanism.

The crater Chang-Ngo at 12.7°S and 2.1°E is about 3 km long and the result of an event with an almost grazing impact. The floor of the crater does not appear uniform, and in the immediate vicinity the floor is saturated with crater pits. The impactor was probably a body of fairly loose consistency that broke into many individual pieces because of gravitational forces, shortly before impact.

The craters Ptolomaeus, Alphonsus and Arzachel undoubtedly form one of the most impressive crater triplets on the Moon's nearside. They are also good examples of the determination of crater ages through the superposition method. The whole region around the three craters has been shaped through secondary impacts from the enormous Imbrium impact (forming the Imbrium Sculpture). Ptolomaeus is the oldest, and Arzachel the youngest, of the three craters. All three craters were created before the impact that formed the Imbrium Basin. The floor of Ptolomaeus has been filled by ejecta from the Imbrium impact, which is why under high solar illumination (at Full Moon), Ptolomaeus does not appear dark (unlike Plato, for example), but as bright as the surrounding highlands. The relatively smooth crater floor of Alphonsus, the furrowed crater walls and the ridge that practically divides the crater in two may all be attributed to the Imbrium impact.

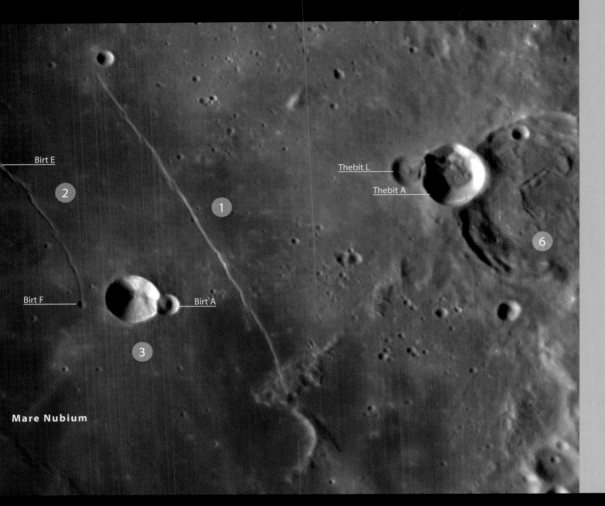

Birt E

2

1

Birt F

Birt A

3

Mare Nubium

Thebit L

Thebit A

6

Rupes Recta 1
22.1°S, 7.8°W

Rupes Recta – also known as the Straight Wall, Straight Scarp or Wall Beta – is a typical lunar undulation, an embankment. With grazing illumination it is a spectacular surface feature, even for observations with small telescopes. Lengths that are given vary between 110 km and 130 km, and the difference in height between the eastern and western sides amount to between 250 and 300 m, with a width of about 2.5 to 3 km. This gives a gradient of less than 10°, so Rupes Recta is rather a gentle slope and not – as the spectacular shadow might imply – a steep escarpment. The slope was probably created by a subsidence of the western side – the crater Birt side – of Mare Nubium, initiated by shock waves from the Imbrium impact and later altered and deformed by the Mare Nubium lava flows.

During the time around First Quarter, Rupes Recta is illuminated by the Sun at a low angle and casts a shadow that is easily visible even in small telescopes. A few days later, when the day/night border (the terminator) is about 20 to 30° away, the feature disappears and may only be made out in large telescopes. At the time of Last Quarter, Rupes Recta appears as a bright line, and a narrow rille (at right angles to Rupes Recta) becomes visible. North of Rupes Recta lies Promontorium Taenarium (19.0°S, 7.0°W). It was given this name by the Polish astronomer Hevelius; he named it after Cape Tainaron (today Matapan) in the Peloponnese. Promontarium Taenarium is an outlier of the southern highlands and stretches out like a peninsula into the lowlands of Mare Nubium.

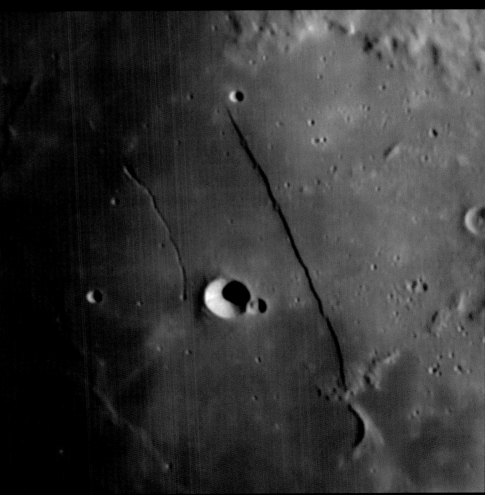

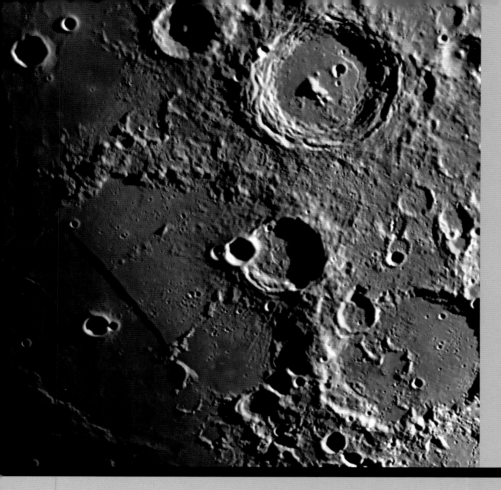

Rima Birt 21.0°S, 9.0°W

Rima Birt, a linear rille, runs for a distance of about 50 km, approximately parallel to Rupes Recta. It widens to a maximum extent of 1.5 km. It is bounded on the south by crater Birt F (3 km) and in the north by Birt E. Birt E lies on a small, elongated and rounded ridge (4.9 km long, 600 m high) and is undoubtedly of volcanic origin. So Rima Birt is probably a lava channel.

If Rupes Recta and Rima Birt are observed under favourable lighting conditions, one gets the impression that they lie in a large, partially lava-flooded crater. On the eastern side is a semicircular crater wall, broken by Thebit, that is completely visible. The western half is suggested by an arcuate structure, similar to a mare wrinkle ridge.

Birt 22.4°S, 8.5°W

A small, circular and funnel-shaped crater, 16 km in diameter, with the craterlet Birt A (6.8 km) breaking the eastern wall. With the illumination near Full Moon, Birt displays a ray system and to the south Birt's ray system becomes mixed with that from the crater Tycho.

Nicollet 21.9°S, 12.5°W ④

A circular crater, 15 km in diameter, with a flat floor and directly west of Birt in Mare Nubium. It lies on a mare wrinkle ridge, which runs in an arc between Nicollet and Birt.

When both craters Birt and Nicollet are illuminated by the Sun, then the same telescopic field of view provides a good comparison between craters with nearly the same diameter; one funnel-shaped and the other with a flat floor.

Lassell 15,5°S, 7,9°W ⑤

Lassell is a circular crater, 23 km in diameter, with flat crater walls and a level floor. West lies Lassell D, a crater pit of just 2 km diameter and 400 m depth. It is surrounded by a small, but extremely bright halo of ejected material, and under steep solar illumination exhibits a weakly developed ray system.

Thebit 22.0°S, 4.0°W ⑥

An interesting crater, 56 km in diameter with terraced walls. The crater's floor is rich in structure. The western wall has been broken by the crater Thebit A (about 20 km) and the northwestern wall of Thebit A has itself been further breached by Thebit L (10 km). Thebit L has a definite central peak, which is unusual for a crater of this diameter. Thebit P (c. 80 km, 24.0°S, 6.0°W), lying southwest of Thebit, is a lava-flooded crater with a partially dark floor. The western portion of the crater is open to the lava flows of Mare Nubium.

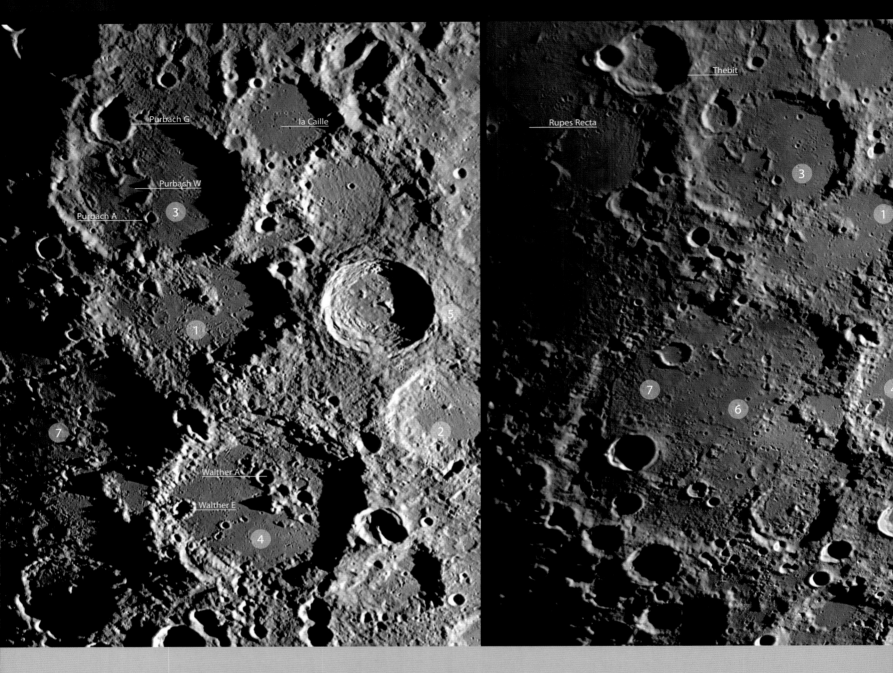

Regiomontanus 28.3°S, 1.0°W   ①

Regiomontanus is an irregularly shaped feature, hardly still recognizable as a complex crater of about 130 × 108 km. The central peak, to the north and lying nearly on the edge of the crater Purbach, towers about 1.2 km above the crater floor. Right on top of the highest peak is a small impact crater. This crater is known as Regiomontanus A, and has a diameter of 5.6 km.

Aliacensis 30.6°S, 5.2°E   ②

A crater, 80 km in diameter, with a smooth, level floor. The small point of a central peak is still detectable. In its morphology, it closely resembles the crater Abulfeda.

Purbach 25.5°S, 1.9°W   ③

One of the smaller complex craters, diameter 115 km, directly adjoining the northern crater wall of Regiomontanus. Its northern wall (reaching just 3 km above the crater floor) is broken by Purbach G (27 km). On the eastern crater floor there are numerous craterlets and right in the centre is Purbach W (20 km), which ap-

pears like a circular flat depression and is heavily eroded. Purbach A, well-preserved, lies directly to its south.

Walther 33.1°S, 1.0°E   ④

Walther is a large and well-preserved complex crater, with a diameter of 138 km, lying directly east of Deslandres. On the eastern floor of the crater is a prominent mountain massif, which is overlapped by larger craters (the largest is Walther A, with a diameter of 12 km). At the western inner wall of the crater lies Walther E (13 km). The original crater name of Walther was, remarkably, allocated by the IAU to a crater pit of just 1 km diameter at 28.0°N and 33.8°W.

Werner 28.0°S, 3.3°E   ⑤

A prominent crater, 70 km in diameter, lying northwest of Aliacensis. The wall of the crater has a sharp crest and, in contrast to Aliacensis, the floor of the crater is rough and furrowed. Multiple central peaks are visible. A massive landslide may be observed within the western wall of the crater.

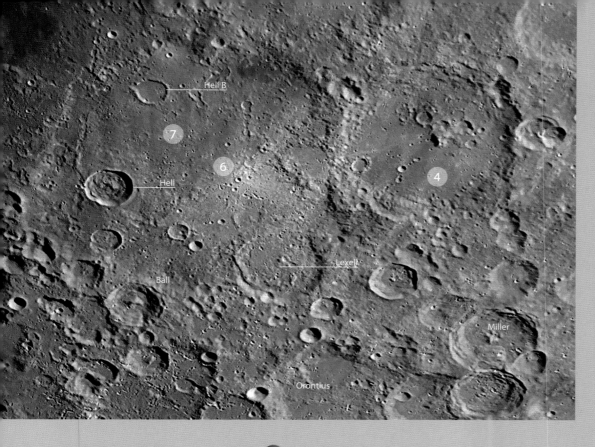

Deslandres
32.5°S, 5.2°W

With its diameter of 256 km, Deslandres is one of the largest craters on the nearside of the Moon. Its crater wall is broken on the south by Lexell and in the east by Walther. On the western floor of Deslandres lies Hell (33 km, 32.4°S, 7.8°W), which has a chaotic structure to its floor. The crater Hell B, diameter 22 km, lies immediately below the northern crater wall of Deslandres. Its floor is, in comparison with Hell, flooded with lava and appears completely smooth and level. Large telescopes reveal a few crater pits (< 1 km). Its northwestern wall has been breached. East of Hell B, against the inner crater wall, lies a conspicuous, dark, irregularly shaped lava surface.

The bright ray system and the dark lava surface in the northeast of Deslandres form a striking contrast. Deslandres formed in the Pre-Nectarian Epoch (between 4.5 and 3.92 billion years ago). Because of its size, Deslandres could also be a small lunar basin, but it lacks any indications of concentric outer ramparts, and there is no measurable gravitational anomaly. If they ever existed, they have been completely destroyed by later impacts and ejecta. Deslandres is so eroded that it was not given its own name in early Moon maps. Such maps named the region the 'Hell plain' after the conspicuous crater Hell.

Cassini's Bright Spot

In the centre of the eastern portion of the floor of Deslandres there lies a prominent chain of five craters, that become progressively smaller in diameter. If one extends the chain of craters slightly towards the south, one comes to a group of smaller craters. The largest of this group has a diameter of about 5 km and is surrounded by a bright halo and an asymmetrical ray system of ejected material (similar to that of Proclus). The system of rays is directed towards the east in the form of a fan, deposited over an angle of about 90°. The longest streak runs directly towards the east and comes to an end only about 50 km away, in the centre of the crater Walter. When illuminated at Full Moon, the small halo around the crater is one of the brightest spots on the Moon's surface and appears almost as a point of light – almost like a bright star.

Cassini, the famous astronomer of the 17th century, claimed to have seen a bright white cloud in this area, leaving a new crater after it had dispersed. The small crater is certainly one of the youngest on the Moon's surface, and is no older than Tycho (about 100 million years old).

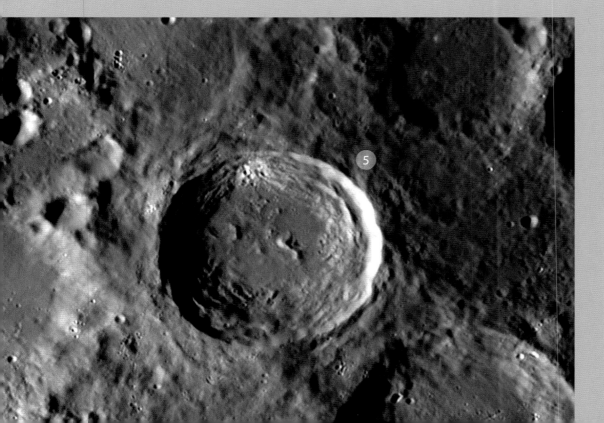

Maurolycus 42.0°S,14.0°E   ①
Faraday 42.4°S, 8.7°E   ②
Stöfler 41.1°S, 6.0°E   ③

These three craters form an impressive triplet for any size of instrument. Maurolycus and Stöfler were classed, on old Moon maps, as walled plains because of their size, and Faraday as a crater. The whole region around these craters in the southern highlands shows how violently and chaotically the Late Heavy Bombardment (LHB), more than 4 billion years ago, restructured the landscape through numerous impacts following one another in time, and superimposed on one another in space.

Maurolycus is an enormous complex crater with a diameter of 114 km, terraced inner walls and a central peak. The crest of the wall lies about 4.7 km above the crater's floor. The extent of the shadows shows that Naurolycus is the deepest of the three craters. The northwestern crater wall has been breached by Maurolycus F, and on the crater's floor there are the smaller craters Maurolycus M, J and A.

Faraday, probably the youngest of the three craters has a diameter of 69 km and a wall height of about 4 km. It lies in the centre of the three craters. The crater wall has been breached by three large, conspicuous craters, with Faraday A on the northeastern wall, and Faraday P and C on the southwestern wall.

Stöfler is a very large complex crater, with a diameter of 126 km and a depth of 2.7 km. Visual observations show a few small craters on the crater's floor, but large instruments reveal dozens of craterlets and pits. The eastern portion of the crater's floor is mountainous and gives the impression of being the remnants of a former crater, which was almost completely destroyed by the Faraday impact. The bright rays that run across Stöfler consist of ejected material from the Tycho impact.

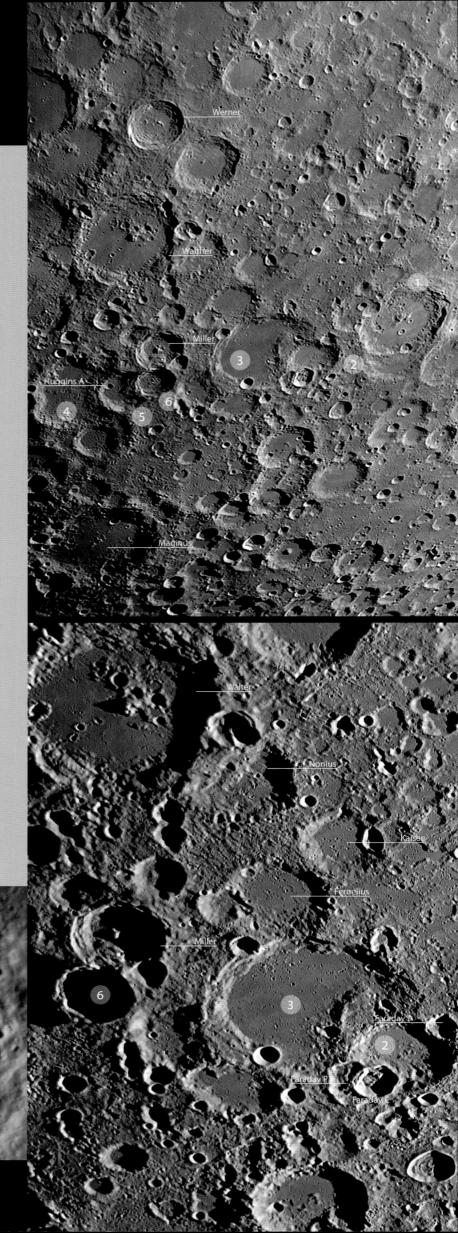

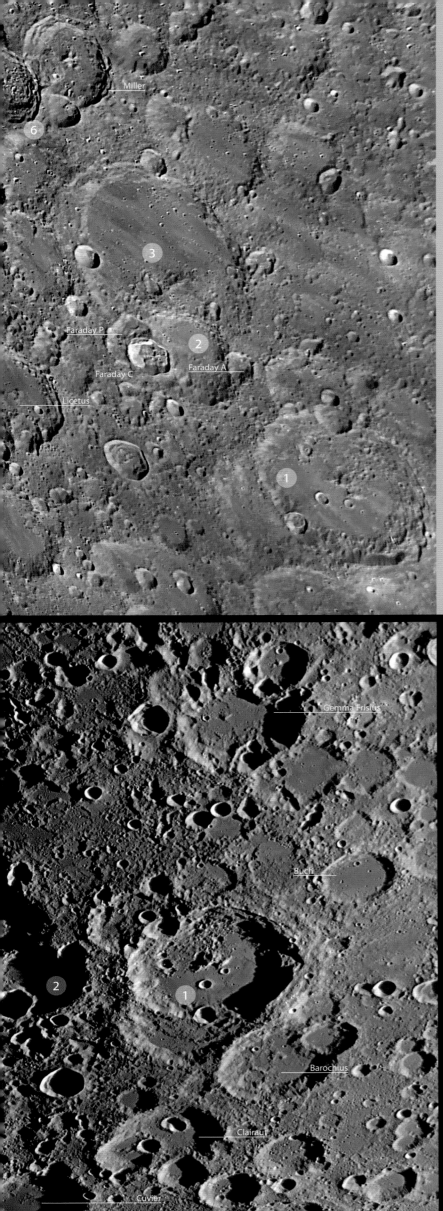

Orontius 40.3°S, 4.0°W   **4**
Huggins 41.1°S, 1.4°W   **5**
Nasireddin 41.0°S, 0.2°E   **6**

A good example for visualizing the superposition method, where the temporal sequence of impacts events may be determined, is shown by the three craters Orontius, Huggins and Nasireddin. The three craters lie directly west of Stöfler. The large crater Orontius (125 km) was formed first. Its eastern wall was overlapped by Huggins (65 km), which was formed by a subsequent impact. The eastern wall of Huggins was destroyed by the smallest of the three craters, Nasireddin (52 km). This gives a succession in time for the impacts: Orontius, Huggins and finally Nasireddin.

The central peak of Nasireddin lies almost exactly on the zero meridian of the lunar coordinate system. Huggins also has a large central massif, and the craterlet Huggins A on the crater floor. The outer, western wall of Huggins lies over a wide area of the crater floor of Orontius.

The sequence in time of the crater Miller (75 km, 39.3°S, 0.8°E) cannot be determined without additional information, because the crater walls of Nasireddin and Miller coincide with one another.

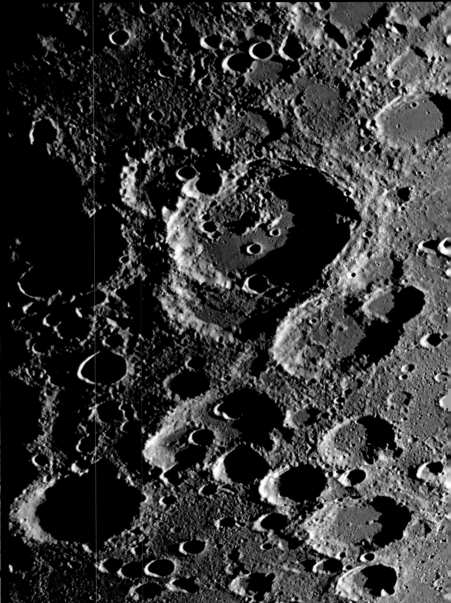

South Pole 90°S ①

The Moon's South Pole lies west of the barely 70-km crater Malapert (84.9°S, 12.9°E) – but slightly farther south. The next crater, lying to the west is Cabeus (84.9°S, 35.5°W), just about 100 km in diameter. The South Pole itself is difficult for even experienced lunar observers to locate, because it lies in an area that is saturated with craters, and the lighting and shadows create a chaotic appearance. Observation of the South Pole therefore requires optimum libration angles.

Moretus 70.6°S, 5.8°W ②
Gruemberger 66.9°S, 10.0°W ③

Moretus is a relatively young crater, 110 km in diamater, conspicuously terraced inner crater walls and a large, pyramid-shaped central mountain. It undoubtedly was created in the Eratosthenian period on the lunar timescale. It is a magnificent sight under favourable libration conditions in both large and small telescopes.

In stark contrast to Moretus is the significantly older, neighbouring crater Gruemberger, lying directly to the northwest, and which, with a diameter of 94 km, is of a similar size to Moretus. It is obvious from its heavily eroded crater wall that it is distinctly older. The crater Gruemberger A (20 km) lies on the crater's floor.

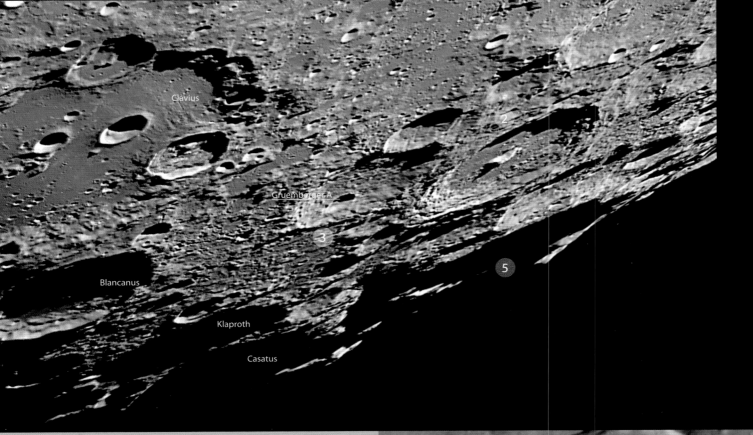

Curtius 67.2°S, 4.4°E **④**

A crater, 95 km in diameter, with an eccentrically placed central peak, which lies northeast of Moretus. Smaller, later impacts have been superimposed on the crater wall in the south and the east. The crater's floor appears smooth and level.

Newton 76.7°S, 16.9°W **⑤**

A crater with a diameter of 78 km and a depth that cannot be precisely measured from Earth because of its extreme position on the limb. Newton is very difficult to observe, because it is right next to the South Pole. In the area near the South Pole, under favourable libration conditions a few mountain peaks of the outer rampart of the South Pole–Aitken Basin may be observed. On historical Moon maps, the name Newton was also given to an elliptical feature, similar to a ghost crater, directly south of Plato.

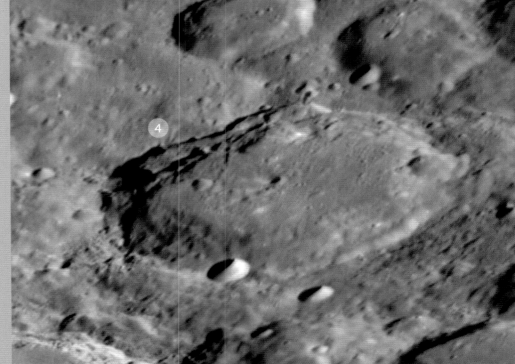

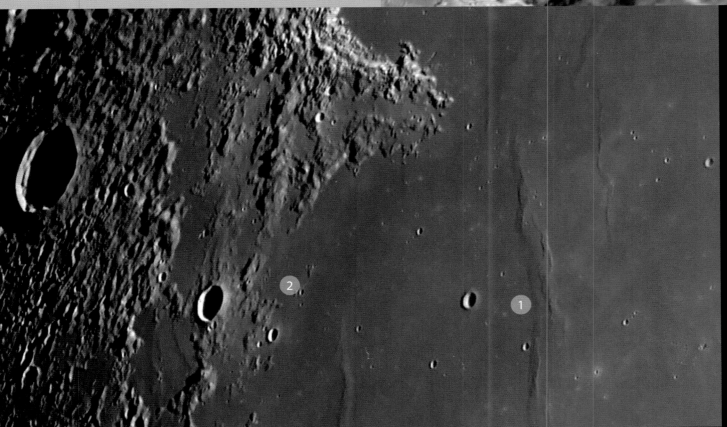

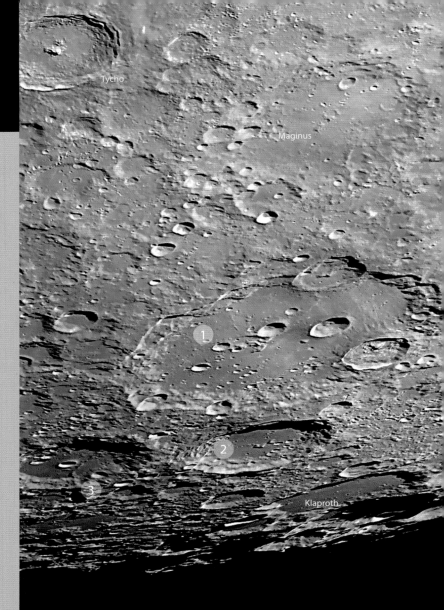

Clavius 58.8°S, 14.1°W ①

Clavius, with its diameter of 245 km and its slightly polygonal shape, is certainly one of the most spectacular complex craters on the nearside of the Moon, and therefore one of the favourite objects for observing with any size of instrument, even in binoculars or a spotting scope. Clavius is suitable as a test object for the determination of a telescope's resolution. The best observation time is shortly before First Quarter (at sunrise) or shortly after Last Quarter (at sunset), but observation is worthwhile at any time that Clavius is illuminated by the Sun. Depending on the angle of incidence of sunlight, different details of the internal structure and its crater walls are revealed. When fully illuminated (at Full Moon) Clavius becomes more or less invisible.

Clavius was formed in the Nectarian period on the lunar timescale. The southern crater wall is broken by the crater Rutherfurd (48 km, 60.9°S, 12.1°W) and the northern wall by Porter (51 km, 56.1°S, 10.1°). Both craters exhibit central mountain massifs. Between Rutherfurd and Porter on the crater floor there are low ridges, like mare wrinkle ridges. Beginning at Rutherfurd, a chain of craters, gradually decreasing in size, stretches in an arc towards the west: Clavius D, then Clavius C, N and J (21 km, 13 km and 12 km, respectively). Clavius D, the largest of these secondary craters has a diameter of about 25 km. On the southern floor of the crater, near the inner crater wall, larger telescopes show dozens of craterlets and crater pits.

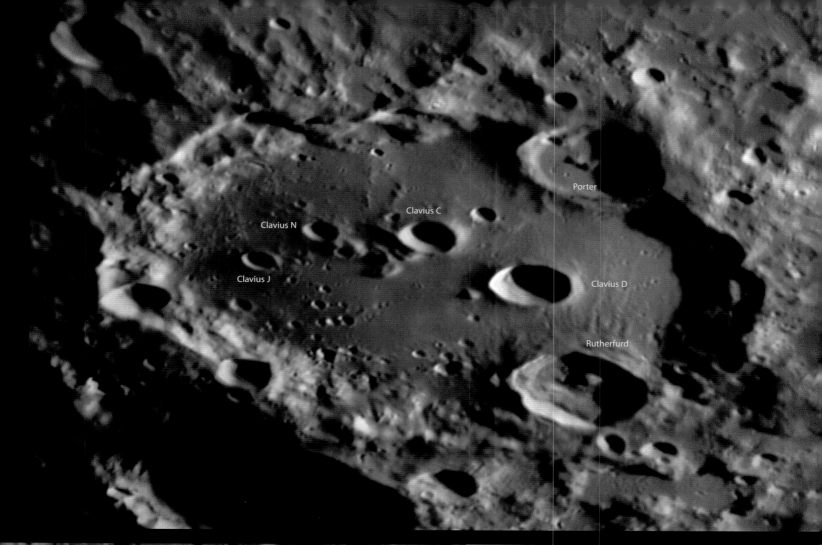

Porter

Clavius C

Clavius N

Clavius J

Clavius D

Rutherfurd

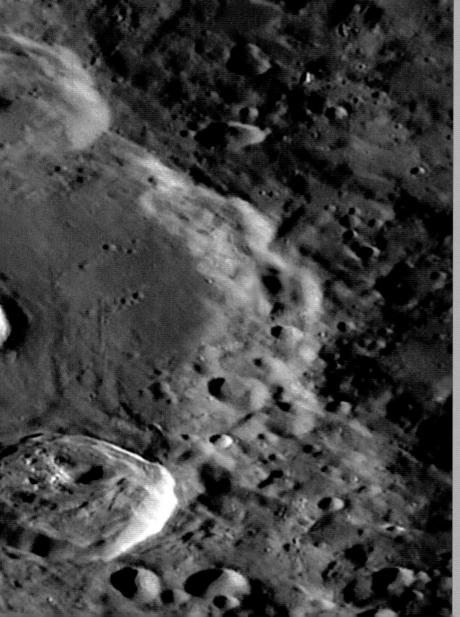

Blancanus 63.8°S, 21.4°W ②

A large crater, 117 km diameter, with a fairly smooth, level floor and a small central peak. Blancanus lies southwest of Clavius, and in the southern portion of the crater's floor there is a group of smaller craters.

Scheiner 60.5°S, 27.5°W ③

A crater without a central peak and a diameter of 110 km. In the centre of the crater's floor lies the small, but clearly recognizable, crater Scheiner A (12 km). Scheiner A may have destroyed the central peak, which must have once existed for a crater of this size. One of the rays of Tycho's ejecta covers the western wall of Scheiner. Under grazing illumination, a low feature, resembling a mare wrinkle ridge, becomes visible, crossing the crater's floor from southwest to northeast, and interrupted by Scheiner A.

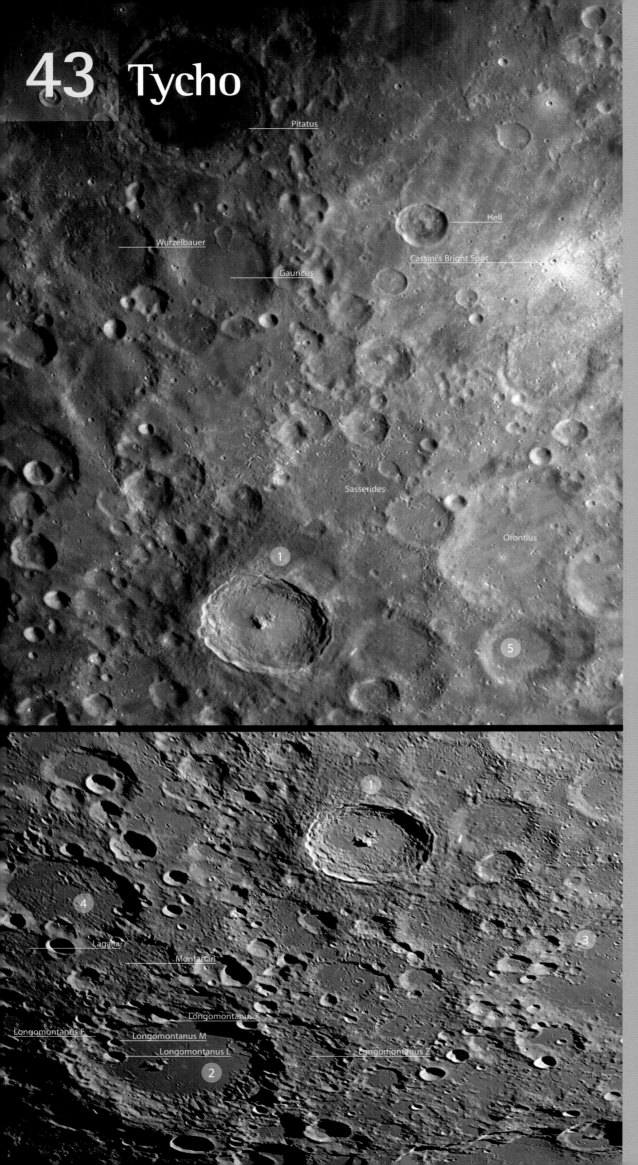

Pitatus

Hell

Wurzelbauer

Cassini's Bright Spot

Gauricus

Sasserides

Orontius

1

5

1

4

Lagalla

Montanari

Longomontanus K

Longomontanus F

Longomontanus M

Longomontanus L

Longomontanus Z

2

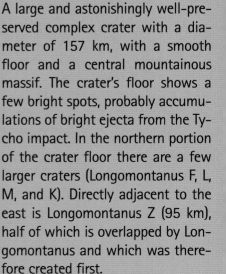

Tycho 43.4°S, 11.1°W ①

Tycho is, with Clavius, undoubtedly the most prominent crater in the southern highlands. Its outer diameter amounts to 102 km and the inner crater wall measures about 85 km in diameter. It exhibits marked terraces on its interior walls and a complex central peak, which appears, from Earth, angular, steep and furrowed. The crater's walls reach a height of 4.8 km above the floor and the central peak towers 2.3 km above that floor.

Under high illumination, Tycho exhibits the largest and brightest of all ray systems. Tycho is certainly one of the youngest lunar craters, probably created only about 100 million years ago. Portions of the ray system may be traced over a distance of 1800 km. The most conspicuous and brightest rays stretch towards the east (Stöfler), southwest (Longomontanus, Clavius) and northwest (Kies, Bullialdus). The ray system is so bright that it is even visible by earthshine. Shortly after First Quarter, Tycho appears like a black chasm, surrounded by thick walls.

Under grazing illumination it becomes obvious that the crater is surrounded by a ring zone of dark material with a diameter of about 150 km. It consists of lunar crust that was melted by the energy of the impact and has subsequently solified.

Longomontanus ② 49.6°S, 21.9°W

A large and astonishingly well-preserved complex crater with a diameter of 157 km, with a smooth floor and a central mountainous massif. The crater's floor shows a few bright spots, probably accumulations of bright ejecta from the Tycho impact. In the northern portion of the crater floor there are a few larger craters (Longomontanus F, L, M, and K). Directly adjacent to the east is Longomontanus Z (95 km), half of which is overlapped by Longomontanus and which was therefore created first.

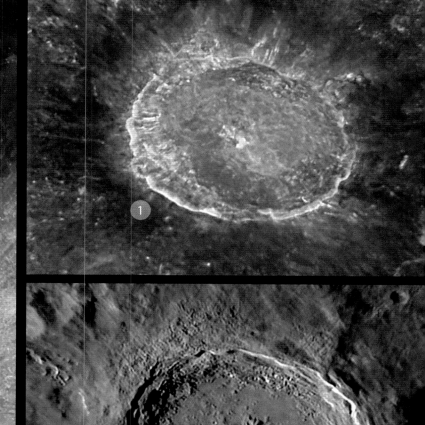

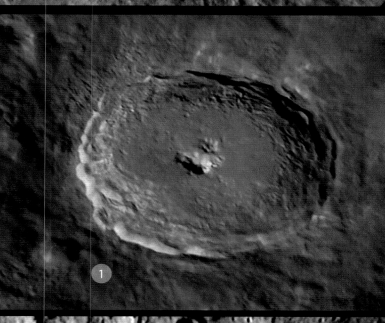

Maginus 50.0°S, 6.2°W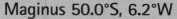

Maginus, with its diameter of 194 km, appears gigantic, and is among the largest complex craters on the Moon's nearside, almost in transition to a small impact basin. The crater's walls and the crater rim are heavily worn away and saturated with smaller craters from later impacts. In the centre of the floor is the remnant of a central massif, also partially covered with craters. Lying on the eastern floor of the crater are the moderately large craters Maginus A and Z, and Maginus F and G on the northwestern crater wall.

Wilhelm 43.4°S, 20.4°W 4

A smaller, moderately eroded complex crater with a diameter of 106 km. The western floor of the crater is fairly smooth and level, whereas the eastern exhibits a few ridges and smaller craters. South of Wilhelm lie two smaller and largely destroyed craters, Montanari (76 km, 45.8°S, 20.5°W) and Lagalla (85 km, 44.6°S, 21.5°W).

Saussure 43.4°S, 3.6°W 5

A crater 54 km in diameter, lying east of Tycho. The crater's floor appears smooth and level, but larger instruments show a few crater pits on its floor.

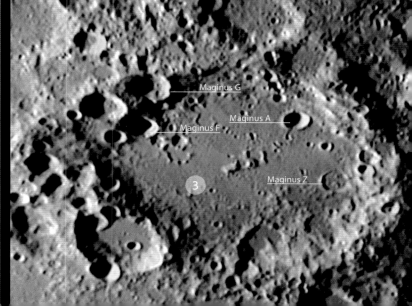

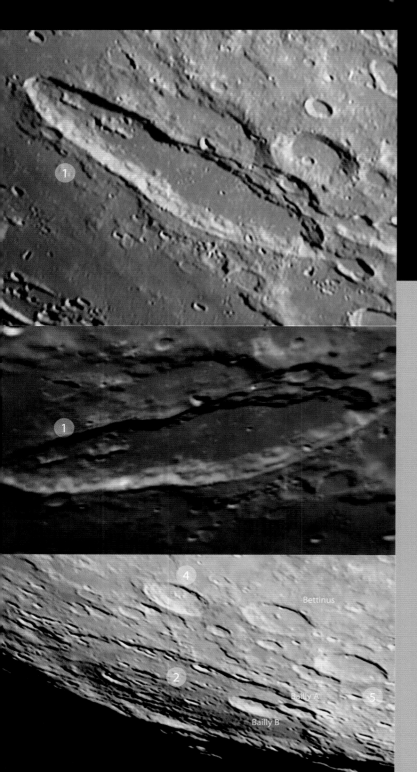

Schiller 51.9°S, 39.0°W

Schiller is one of the few truly elliptical craters, having measurements of 70 × 180 km. The southern portion of the crater's floor is smooth and level, the northern portion is rough and furrowed and has two mountain peaks. Schiller probably once consisted of two or more overlapping craters. NASA experiments in the 1960s have shown that such a structure may result from a further impact that occurs at a very low angle (of just a few degrees), and may perfectly well lead to a form of crater like Schiller.

Schiller–Zucchius Basin

The region between the craters Schiller and Zucchius is a multi-ring basin with two, or possibly three, concentric rings of mountain chains (basin ramparts). The inner ring, about 200 km in diameter, cuts the crater Segner, and the second ring, about 330 km in diameter, cuts across the crater Zucchius. Like many other basins, the Schiller-Zucchius Basin also shows a gravity anomaly (a mascon).

Bailly 66.5°S, 69.1°W

Bailly is the largest crater that may be observed on the nearside of the Moon (and is described on older Moon maps as a walled plain). Nevertheless, because of its extreme position on the limb it is not very conspicuous, and because of the curvature of perspective at the limb is distorted into an extremely elliptical shape. Bailly has a diameter of nearly 300 km, and as such is classified as a medium-size impact basin. The floor of the crater exhibits many craterlets. The two largest are Bailly A (38 km) and Bailly B (65 km). Many of the smaller craters in Bailly exhibit ray systems, and under grazing illumination a few lunar domes may be observed on the crater's floor. Its observation requires favourable libration angles.

Segner 58.9°S, 48.3°W ③

A shallow crater, 67 km in diameter, and with a wavy, rough crater floor. Segner H, a crater 7 km across, lies on the northern portion of the floor, surrounded by a dark halo. It is a Dark Halo Crater.

Zucchius 61.4°S, 50.3°W ④

Zucchius is a crater with a diameter of 64 km, a depth of 3.2 km and with terraced inner walls. Under Full-Moon illumination, Zucchius exhibits a ray system.

Kircher 67.1°S, 45.3°W ⑤

A lava-flooder crater, 72 km in diameter, with a smooth crater floor. Kircher, together with the craters Bettinus (71 km, 63.4°S, 44.8°W) and Zucchius forms a triplet, which is particularly conspicuous when Bailly is still in shadow.

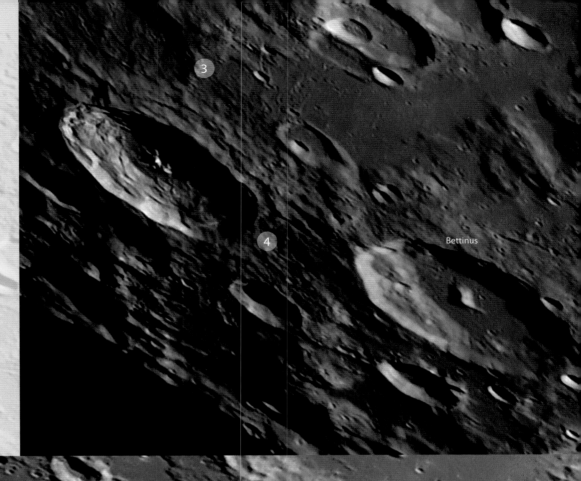

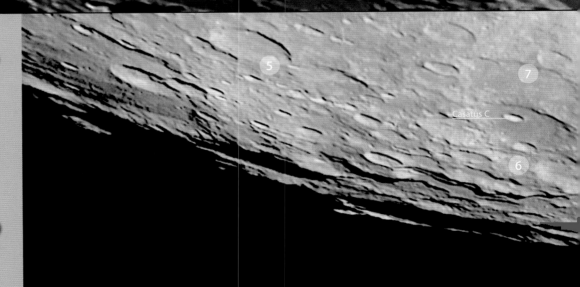

Casatus 72.8°S, 29.5°W ⑥

Casatus is a large, lava-flooded crater with a diameter of 108 km. The western crater wall is higher than the rest of the wall. On the crater's floor lies a smaller crater with a sharp crest, Casatus C (17 km), and the southern wall of Casatus is broken by Casatus J (22 km).

Klaproth 69.8°S, 26.0°W ⑦

A large complex crater, 120 km in diameter, overlying the crater Casatus.

5 Palus Epidemiarum

Palus Epidemiarum ❶
32.0°S, 28.2°W
Lacus Timoris ❷
38.8°S, 27.3°W
The 'Marsh of Epidemics' and the 'Lake of Fear' are irregularly shaped lava plains, which are about 290 km and 120 km across, respectively. Palus Epidemiarum is crossed by numerous rille systems, and Lacus Timoris is surrounded by mountain massifs.

Capuanus 34.1°S, 26.7°W ❸
A lava-filled crater, 59 km in diameter. The crater floor is relatively smooth and exhibits a low ridge. In large telescopes, and with grazing illumination, two large volcanic domes are visible, one of which has a summit crater.

Mare Nubium

5

4

Rupes Mercator

Rimae Hippalus

Rima Hesiodus

9

1

Rimae Ramsden

3

8

2

7

6

Schiller

Mercator 29.3°S, 26.1°W ❹
Campanus 28.0°S, 27.8°W ❺
Two conspicuous craters whose outer walls are in direct contact and with nearly the same diameter (46 km and 48 km, respectively). Campanus has a small curved central peak. Two sections of a nondescript, but easily visible, rille run north-south between the crater walls. Rupes Mercator (30.0°S, 23.0°W), an escarpment about 180 km long, extends from the eastern wall of Mercator. A nameless rille runs away to the southeast.

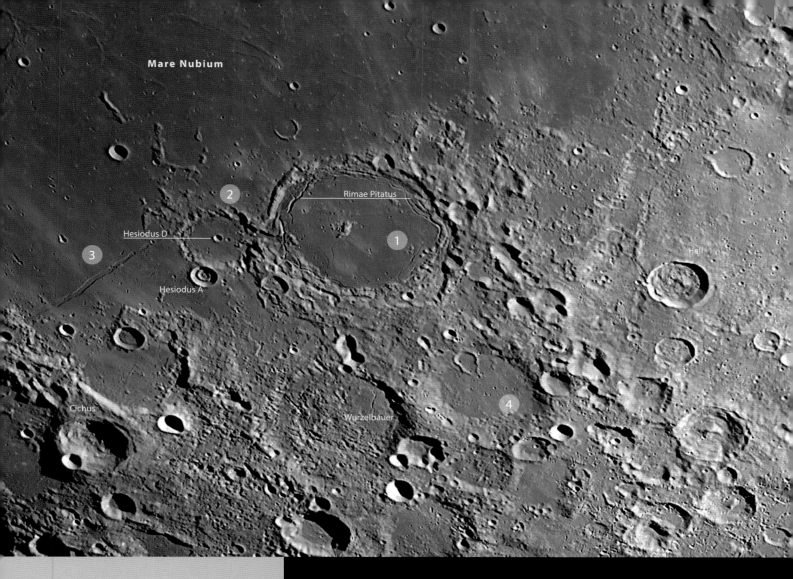

Mare Nubium

Rimae Pitatus

Hesiodus D

Hesiodus A

Cichus

Wurzelbauer

Hell

Rima Hesiodus ③
31.0°S, 22.3°W
A wide (about 3 km) linear rille, with a length of about 250 km. The rille probably arose through through a collapse of the surface. It is clearly visible in smaller telescopes.

Gauricus ④
33.8°S, 12.6°W
A heavily eroded crater, 79 km in diameter. A circular chain of smaller craters lies on the wide western wall. Some overlapping smaller craters lie on the northern portion of the crater floor.

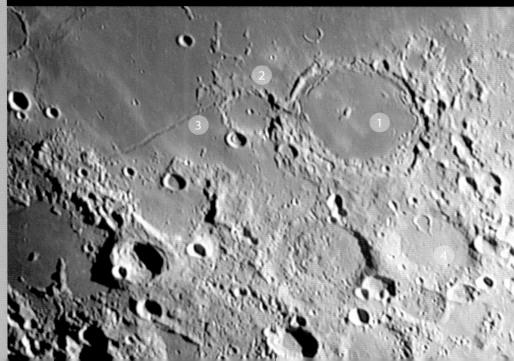

Guericke

Mare Cognitum

Lassell

Arzachel

4

Rupes Recta

2

5

Nicollet

3

6

1

Pitatus

7

8

Mare Humorum

Mercator

Campanus

Palus Epidemiarum

Tycho

Mare Nubium 20.0°S, 15.0°W　①

Mare Nubium ('Sea of Clouds') is irregular, roughly round in shape, with no distinct boundaries anywhere, such as those found with other maria, typically in the form of sections of wall. It is possibly part of the enormous lava flow to which Mare Cognitum, Mare Insularum and parts of Oceanus Procellarum belong. It has a diameter of roughly 750 km and the lava covers an area of about 250 000 km². Mare Nubium offers a multitude of interesting surface features for both visual and photographic observation.

Lubiniezky 17.8°S, 23.8°W　②

An almost completely lava-filled crater, 44 km in diameter, similar to Keis, lying northwest of Bullialdus. The southeastern crater wall is open to the lava surface of Mare Nubium. The maximum difference in height between the crater's wall and the floor amounts to just 770 m. Northwest of Lubiniezky lie the two craters Lubiniezky A and E. The former crater wall of Lubiniezky E (c. 40 km, 16.5°S, 27.2°W) has been practically completely submerged by lava, only a few individual peaks stick up above the surface of the lava.

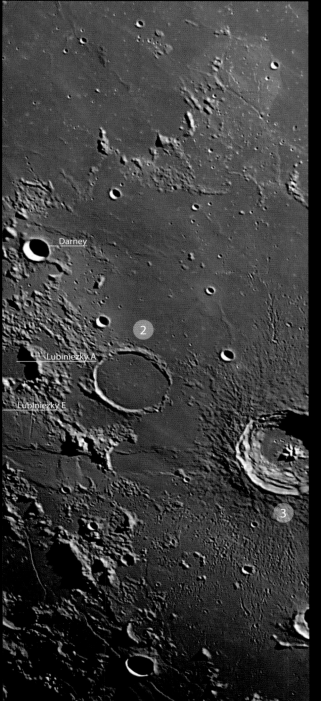

Darney

Lubiniezky A

Lubiniezky E

Bullialdus 20.7°S, 22.2°W ③

A very conspicuous crater, 60 km in diameter, with terraced inner crater walls and a central mountain with multiple peaks. Melted ejecta material lies radially around the outside of the crater. It appears like a smaller version of the large crater Copernicus, but under high solar illumination shows no sign of a ray system, despite the fact that it must be relatively young.

Opelt 16.3°S, 17.5°W ④

Opelt is the remnant of a lava-flooded crater, 48 km in diameter. It appears almost like a ghost crater. The northern crater wall borders a slightly domed, low plateau (possibly a megadome) with a diameter of about 60 km. The plateau is crossed by Rimae Opelt, which has a length of about 70 km. A few small, low ridges, as well as craterlets, crater pits and depressions.

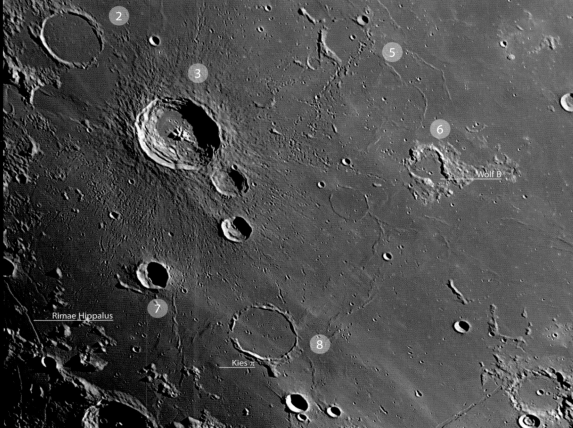

Wolf B

Rimae Hippalus

Kies π

Gould 19.2°S, 17.2°W ⑤

A crater remnant, lying south of Opelt. Gould appears even more eroded than Opelt.

Wolf 22.7°S, 16.6°W ⑥

The heavily eroded remnant of a lava-flooded crater with a diameter of 25 km. It lies on a very low plateau. The remnants of the southern wall are overlapped by the crater Wolf B. Immediately adjacent to the west is a mountain 17 × 17 km.

König 24.1°S, 24.5°W ⑦

A crater, 24 km in diameter. The crater walls rise about 2.4 km above the crater's floor. Bright ejecta from the Tycho impact lie on part of the crater wall, and a section of one of the rays crosses König's floor.

Kies 26.3°S, 22.5°W ⑧

An interesting, almost completely lava-filled crater with a spike-like extension to the southern crater wall. The floor appears smooth, but large telescopes reveal craterlets on the floor. Directly west lies the volcanic dome Kies π, with a diameter of about 12 km. In large telescopes Kies π exhibits a summit crater about 1 km across and, under favourable lighting conditions, depressions on its sides.

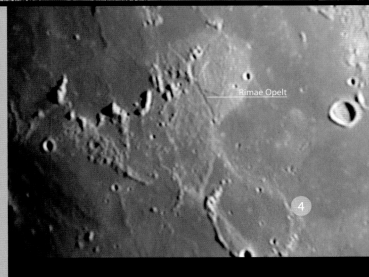

Rimae Opelt

Fra Mauro 6.1°S, 17.0°W

Fra Mauro is a greatly destroyed complex crater, 101 km in diameter. The western half of the crater's floor is entirely covered with hummocks and ejecta from the Imbrium impact. Cracks and graben run across the centre in the north-south direction. The rock samples that the Apollo 14 astronauts collected in the vicinity date the Imbrium impact to about 3.85 billion years ago.

The Apollo 14 landing site lies a short distance north of the crater wall of Fra Mauro. The large rille that runs right through the crater remnant in the north-south direction is a section of the Rimae Parry rille system. The crater is named after a Venetian monk and cartographer, who died in 1459.

North of Fra Mauro lie two north-south oriented ridges; Fra Mauro Eta (20 × 10 km with a summit crater) and Fra Mauro Zeta. Both may be ascribed to the Imbrium Sculpture, and where Fra Mauro Eta may be considered as a small 'pointer' directly indicating the centre of the Imbrium impact.

Gambart 1.0°N, 15.2°W

A circular, lava-filled crater, 25 km in diameter. Gambart A (12 km) west of Gambart is a smaller crater with a ray system. A volcanic dome with a base diameter of about 10 km lies near the craters Gambart B and C (both about 12 km). Marked with an arrow on the image is an object, which might actually be a dome. For its diameter, Gambart is extremely shallow. There are numerous craters in the vicinity with similar morphology, and these have been grouped together as a class known as Gambart craters.

Turner 1.4°S, 13.2°W

A small crater, 11 km in diameter, with a depth of about 2.6 km, which is unusual for such a diameter.

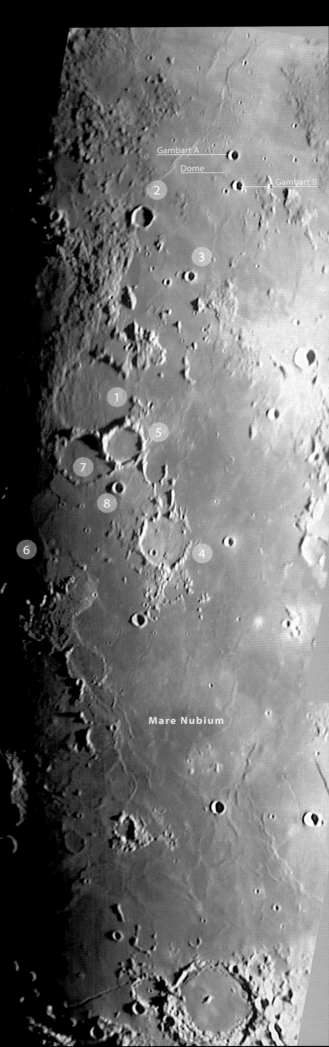

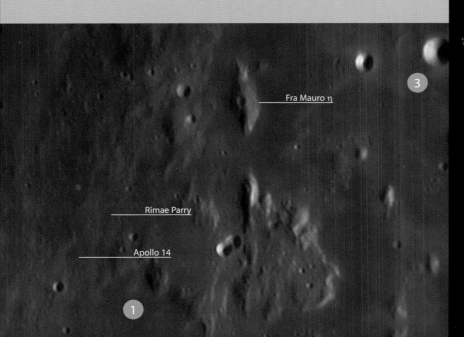

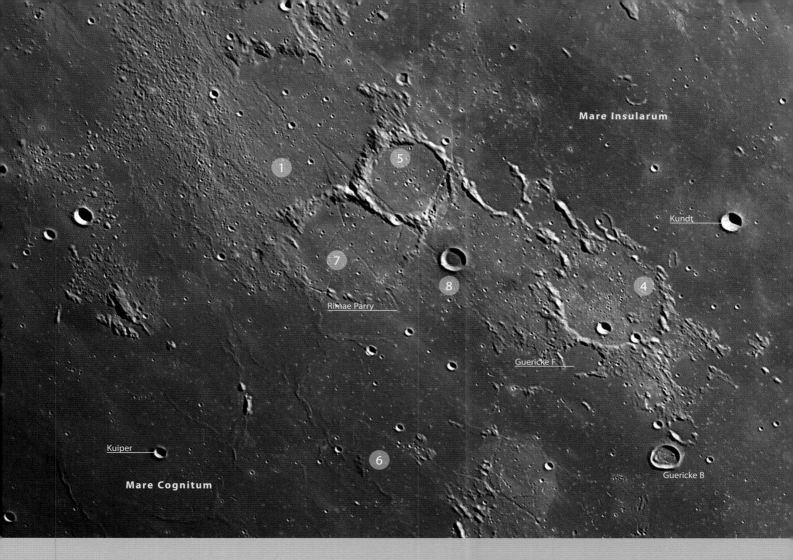

Mare Insularum

Kundt

Kuiper

Rimae Parry

Guericke F

Mare Cognitum

Guericke B

Guericke 11.5°S, 14.1°W
Parry 7.9°S, 15.8°W

Guericke is a heavily eroded remnant of a crater, 63 km in diameter. Adjoining the southwestern wall is Guericke F, with a diameter of about 21 km. The crater wall is open to Mare Nubium.

Parry, 47 km in diameter, lies to the north of Guericke and is also a heavily damaged and lava-flooded crater. A complex system of clefts and rilles, Rima Parry, runs across Parry, Bonpland and Fra Mauro. The system of clefts extends for a length of about 300 km, and is partially visible in even a small telescope.

Mons Moro 12.0°S, 19.7°W

A low ridge, about 10 km long and 3 km wide, lying at the northen end of a mare wrinkle ridge.

Bonpland 8.3°S, 17.4°W

Bonpland, lying southwest of Parry, is the remnant of a shallow crater with a diameter of 60 km. One of the principal rilles of Rimae Parry runs across the floor, breaks through the northen crater wall and ends at Fra Mauro.

Tolansky 9.5°S, 16.0°W

Tolansky is a small crater, 13 km in diameter, lying midway between Guericke and Parry. The crater floor is smooth and, under very low illumination, appears to be slightly domed.

The whole Fra-Mauro region exhibits an abundance of interesting surface features in both large and small apertures. Because most of them are shallow, observation near the terminator under grazing illumination is recommended.

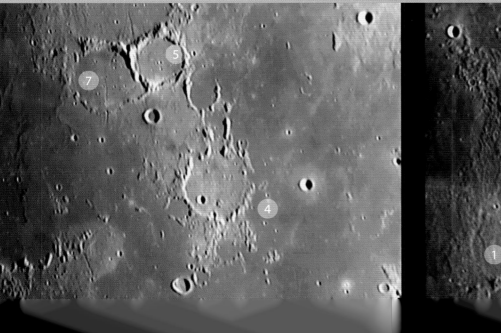

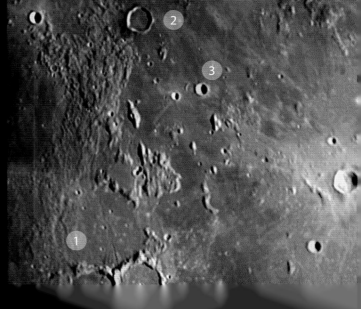

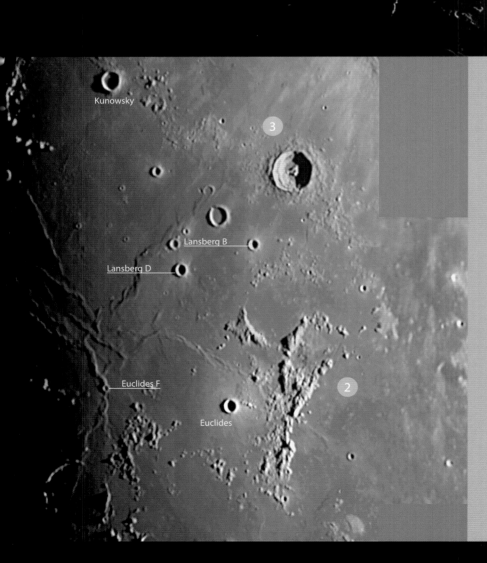

Oceanus
Procellarum

3

2

1

Ranger 7

Mare Nubium

Herigonius

Bullialdus

Kunowsky

3

Lansberg B

Lansberg D

Euclides F

Euclides

2

Mare Cognitum 10.0°S, 23.0°W

Mare Cognitum (the 'Known Sea') has a diameter of approximately 370 km and grades without any transition into Oceanus Procellarum in the west. The basin probably arose in the early Imbrium period of the lunar timescale, and it was flooded by lava in the late Imbrium period. In the northwest it is bordered by Montes Riphaeus (probably the remnants of a large flooded crater) and in the south and southeast by Mare Humorum and Mare Nubium. Until 1964 it was considered to be part of Mare Nubium.

On 31 July 1964, the American lunar probe Ranger 7 crashed southwest of the crater Bonpland and transmitted images until 0.2 seconds before the impact (a total of 4308 images). They were the first close-up images of the Moon's surface, with an image resolution of just about 0.5 m. The mission ushered in a new stage in lunar research. It showed a lava plain saturated with craterlets and crater pits, whereas observations from Earth did not show any craters in this area. As a reminder of this occasion, the name Mare Cognitum was introduced officially into lunar nomenclature by the IAU in 1964.

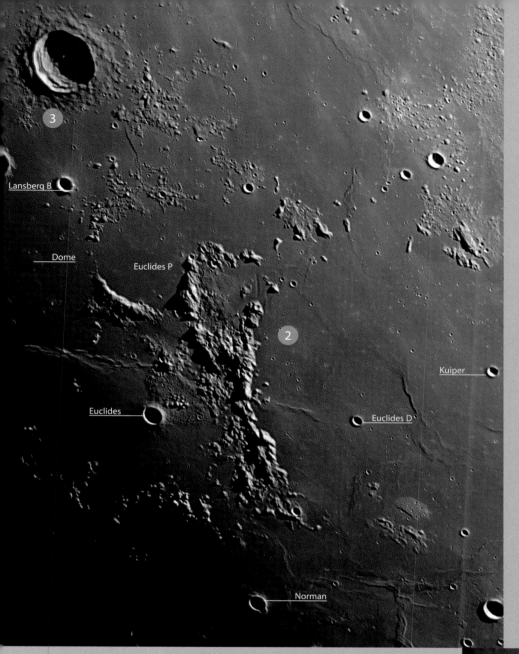

Montes Riphaeus   ❷
7.7°S, 28.1°W

The Riphaean Mountains lie between Oceanus Procellarum (to the west) and Mare Cognitum (in the east). They were named after a chain of mountains in the Urals. They form a chain of mountains about 190 km long, which divide at the northern end into a 'Y'-shape. The highest peaks rise up to 1 km above the surrounding landscape. The extension towards the northeast ends in a large trough, 40 km long. The western portion turns into a semicircular mountain formation, which is probably the remnant of a crater flooded by lava, designated Euclides P. West of the chain of mountains lies the prominent circular, and very bright crater Euclides (7.4°S, 9.5°W), with a diameter of 11 km. Euclides is the centre of a small ray system. Farther north lie two craters, Lansberg D and B (both about 7 to 8 km across), both also very bright and the centres of small, bright ray systems.

Lansberg 0.3°S, 26.6°W   ❸

Lansberg, with a diameter of 38 km, is a prominent crater in this region of the Moon. It has terraced inner crater walls and a central mountain with several peaks. The crater wall rises as much as 3.1 km above the crater's floor. Southwest, and when near the terminator, a very complex, nameless, mare wrinkle ridge system is visible. On one of its sections, the small crater Euclides F (5 km) lies right in the centre. Extending the line of the remnant of the western wall of Euclides P takes one to a large lunar dome (near, and to the southeast of Lansberg D, 10 km).

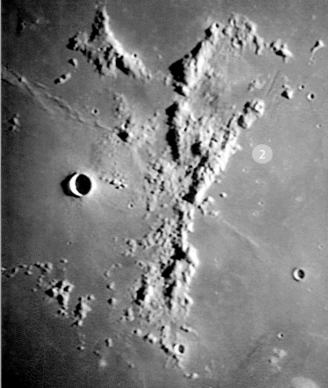

50 Mare Insularum

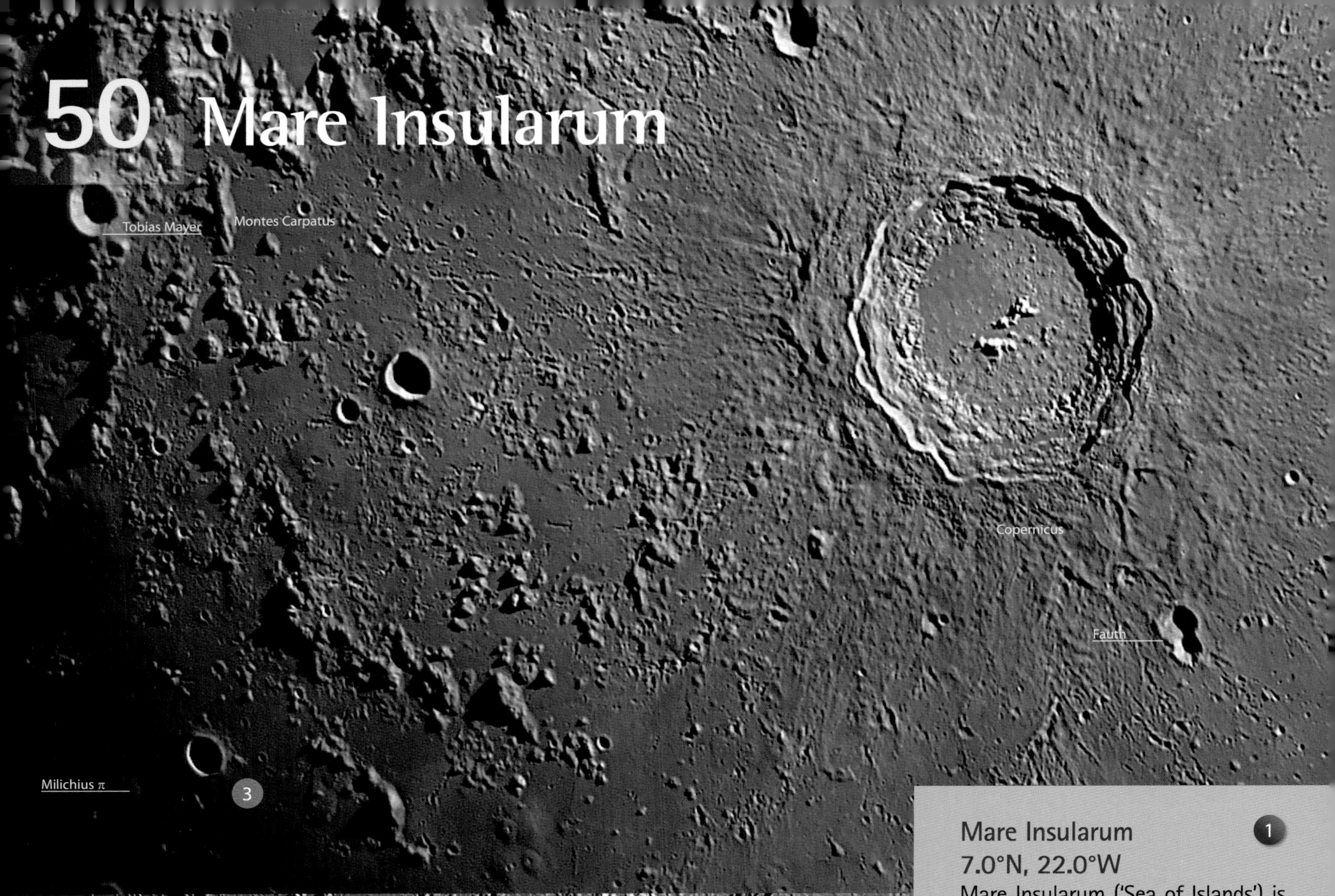

Tobias Mayer
Montes Carpatus

Copernicus

Fauth

Milichius π

3

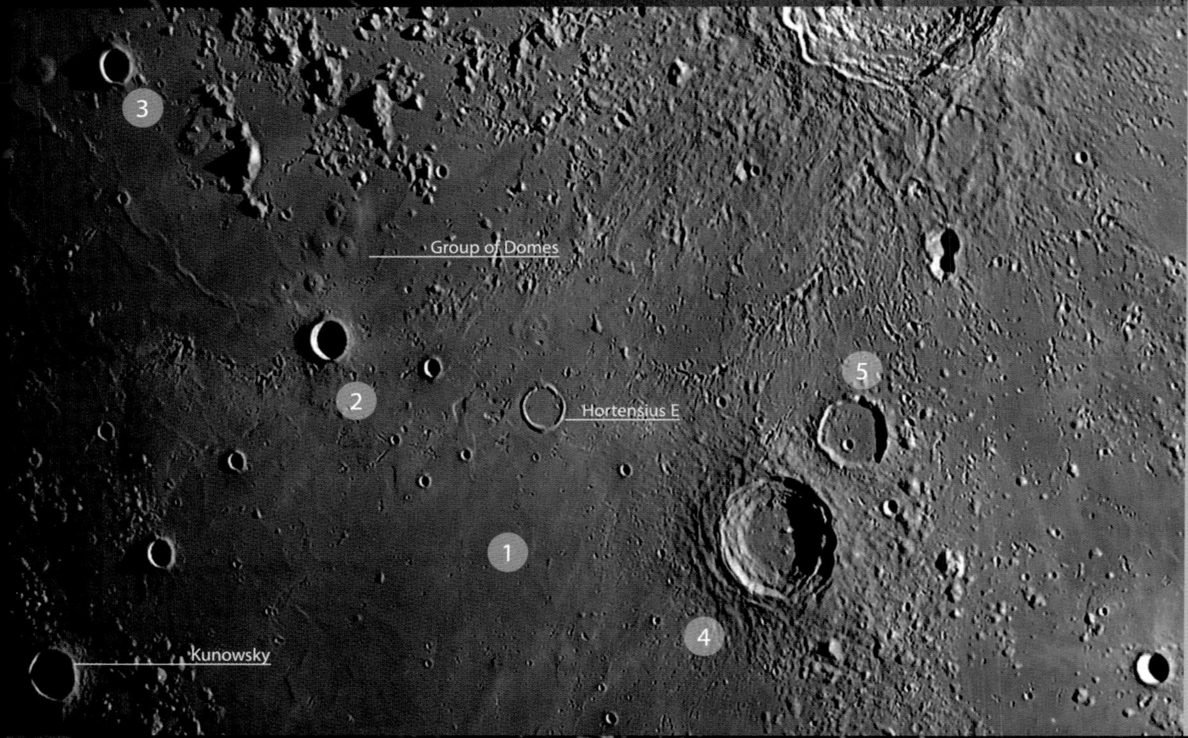

3

Group of Domes

5

2

Hortensius E

1

4

Kunowsky

Mare Insularum 1
7.0°N, 22.0°W

Mare Insularum ('Sea of Islands') is a name first officially introduced by the IAU in 1976, and is therefore not found on older Moon maps. Mare Insularum covers an area of about 900 km in diameter. The centre lies roughly between the two crater Kepler and Encke (on the west) and Sinus Aestuum (on the east). The boundary in the north is formed by the Carpathians (Montes Carpatus). The southern boundary is indeterminate and here Mare Insularum merges into Mare Cognitum (another one of the IAU's recent names). The very young crater Copernicus (formed about 800 million years ago) and its bright ray system lie, like an island, in the 'Sea of Islands'.

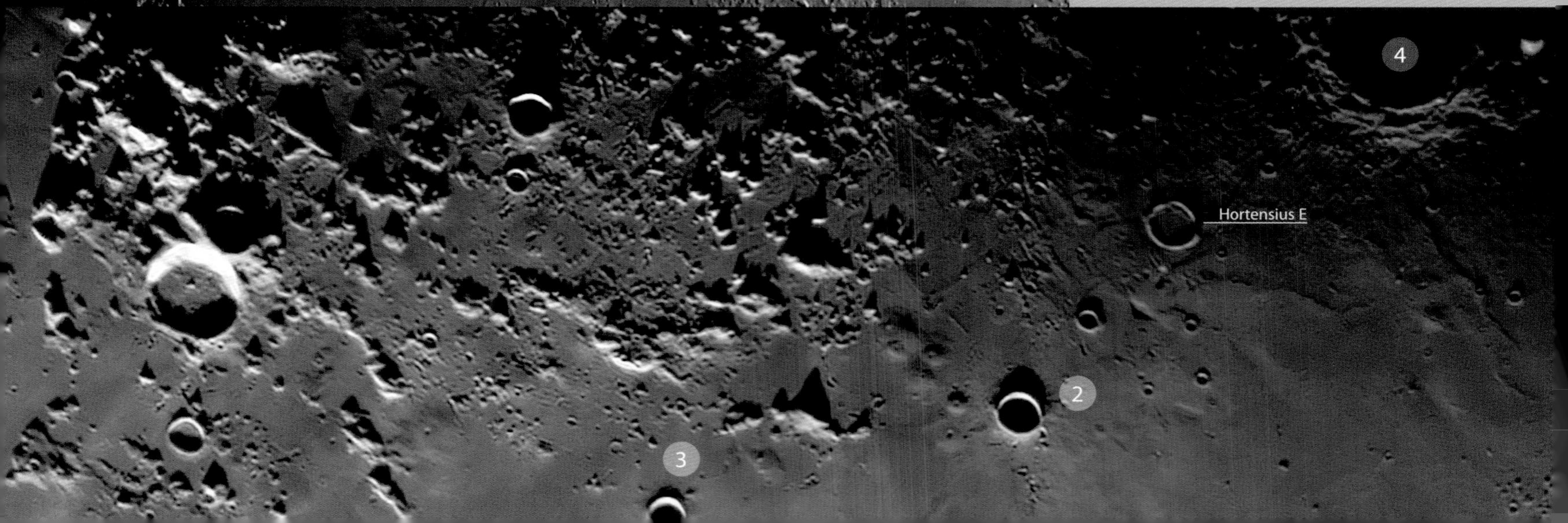

4

Hortensius E

2

3

Hortensius 6.5°N, 28.0°W ②
Milichius 10.0°N, 30.2°W ③

Two circular craters lying west of Copernicus, with diameters of 14.5 km and 13 km, respectively. They are primarily of interesting in that they serve as landmarks for the visual observation or photography of one of the most interesting classes of lunar features, that of lunar domes. The observation of these structures (whose height normally amounts to just a few hundred metres) always requires illumination near the terminator, that is, a low angle of incidence of sunlight.

In this region of the Moon, individual domes and groups of domes may be readily observed. The are the largest lunar domes on the nearside of the Moon. In their form, gradient of their slopes and formation they may most readily be likened to terrestrial shield volcanoes. Directly west of Milichius is the dome Milichius Pi. North of Hortensium lies a group of six such domes. The diameters of indivicual structures lie between 5 and 10 km, but their heights, on the other hand, only amount to a few hundred metres. The gradients of the slopes are extremely low. The summits of many of the domes have caldera-like depressions (with summit craters about 0.9 to 1.5 km in diameter), and a very few even have two. Additional domes also lie in the area between Milichius and the crater Tobias Mayer (33 km, 15.6°N, 29.1°W), as groups and individual, isolated domes. A large megadome plateau lies south of T. Mayer.

The summit craters of the domes offer an excellent means of testing the performance of individual instruments. At apogee, 1 km on the Moon corresponds to 0.51 arcsecond and, at perigee, 0.57 arcsecond. For smaller telescopes, 5 km on the Moon corresponds to 2.55" at apogee, and 2.85" at perigee. Local seeing conditions during observation may also be evaluated in this way.

Southwest of Milichius lies Rima Milichius, a sinuous rille, which is, however, visible only in very large telescopes.

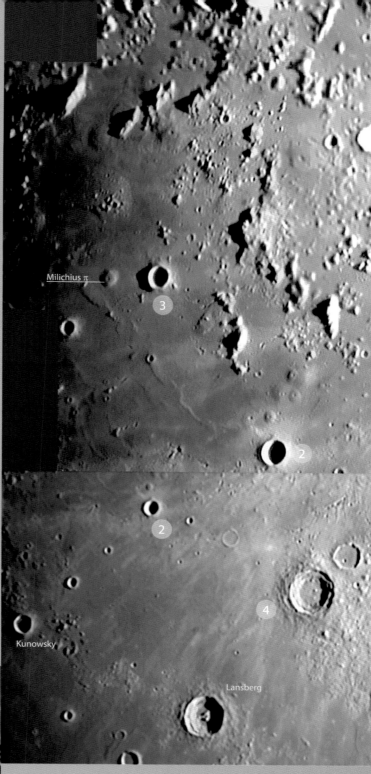

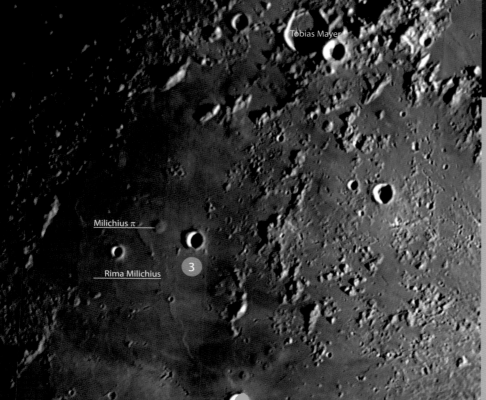

Reinhold 3.3°N, 22.8°W ④
A prominent crater, with a diameter of 42 km, terraced inner walls and a small central peak. A volcanic dome with a caldera-like summit crater lies southwest of Reinhold.

Reinhold B 4.3°N, 21.7°W ⑤
A lava-filled crater, 26 km in diameter, with just a low crater wall and a crater pit on the floor. Although it lies directly next to Reinhold, the interior of the crater is completely different in its structure to that of Reinhold.

51a Copernicus

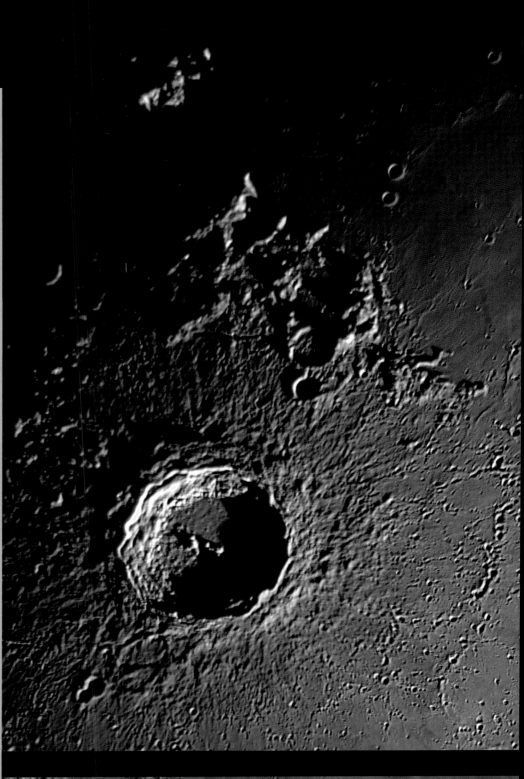

Copernicus 9.7°N, 20.1°W ①

Copernicus is perhaps the most impressive example of a young, large complex crater, and which was formed about 800 million years ago. It lies between the southern side of Mare Imbrium and the northern portion of Mare Insularum, which has no distinct boundaries and merges into Oceanus Procellarum and Mare Nubium.

Copernicus has a diameter of 93 km and distinctly terraced inner crater walls, which tower 3.7 km above the crater's floor. A group of central mountains reaches up to heights of 1.2 km above the floor. The height of the outer ramparts amounts to only about 900 m. The southern portion of the crater's floor is saturated with a vast number of small hills, but the northern portion, by contrast, seems extremely smooth, level and structureless. It has probably been flooded by lava. The ratio of its depth to diameter of 1:25 (known as Schröter's Rule) corresponds more closely to that of a flat plate, rather than to the section of a bowl or funnel. When the Moon's age is about 9 to 10 days, a small lunar dome becomes visible about 90 km west of Copernicus. A large number of smaller secondary craters, crater pits and depressions lie in the terrain to the northeast of Copernicus. They were created by secondary impacts. A bright, extensive and very complex ray system becomes visible around the crater under high solar illumination.

Gay-Lussac

Eratosthenes

Stadius

Hortensius

Mare Insularum

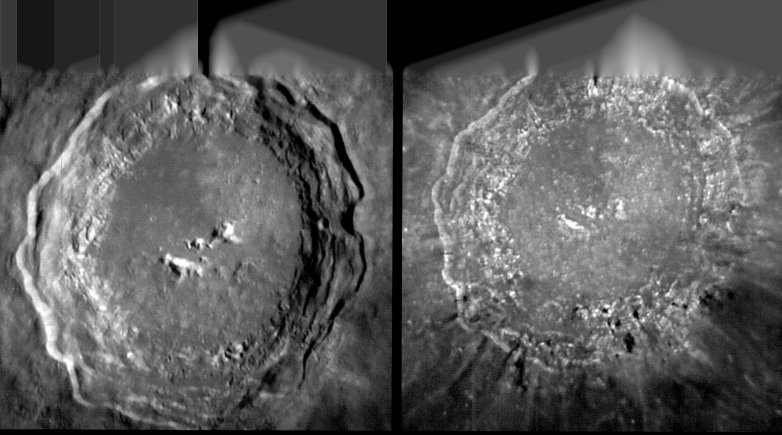

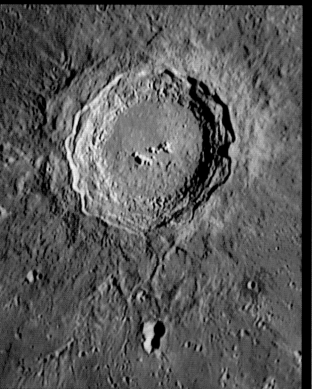

The terraces around the inner walls appear six-sided and in sections. The number of visible central peaks depends on the experience of the observer and the size of the instrument used. Three of the peaks are, because of their height, also visible in small telescopes. Larger telescopes show significantly more mountain peaks. The central mountains contain large quantities of olivine, which was actually formed deep beneath the surface of the Moon, but which was brought to the surface by the enormous impact.

The radial scratches, furrows and notches around the crater's rim were caused by ejecta that was melted by the impact, and which are still easy to see because of the crater's young age. In comparison to similarly large, but significantly older craters in the southern highlands, Copernicus is 'fresh', as if it had just been created. Landslides are clearly visible around the eastern inner crater wall, possibly created by shock waves from the much later Tycho impact.

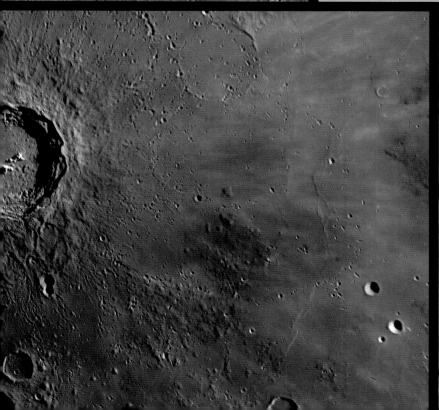

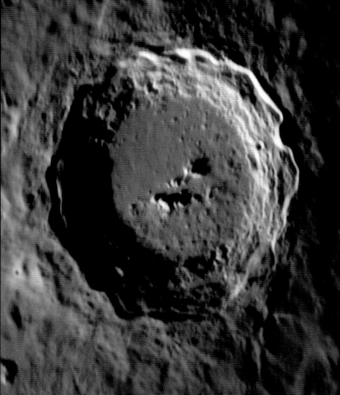

b Copernicus

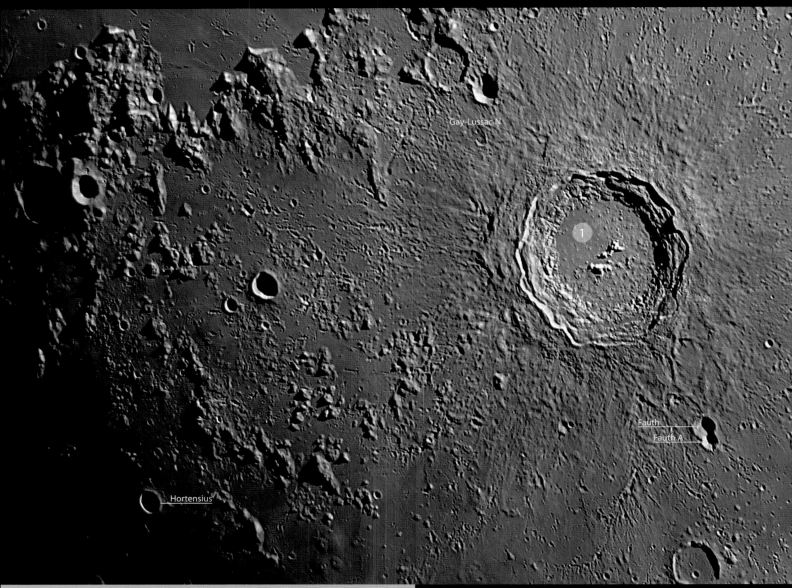

Close to the edge of the crater are two small craters: Copernicus H with a diameter of 5 km (to the southeast) and Gay-Lussac N, with a diameter of only 2 km (to the north). Both craters are surrounded by dark haloes and lie as if stamped into the Copernicus ejecta. Multispectral investigations show that the dark haloes consist of pulverized mare basalt, which was excavated from beneath the ejecta from Copernicus. Both craters are, admittedly, small, but they must be deep enough to have brought the basalt to the surface. They are included in the class of Dark Halo Craters. Two craters that arose similarly, and with dark haloes of mare basalt, lie southeast of Theophilus. Craters with dark haloes – like the bright ray systems – are best seen under high solar illumination. Because they are generally very small, it is advisable to use magnifications of 150 to 200× for visual observation.

Two smaller, overlapping craters, Fauth (12 km, 6.3°N, 20.1°W) and Fauth A (10 km) lie directly south of Copernicus. Fauth and Fauth A are most probably secondary craters, originating in the gigantic Imbrium impact. The two craters resemble a keyhole, and when taken together with Copernicus, the three craters serve as an indicator of the north–south direction of the lunar coordinate system.

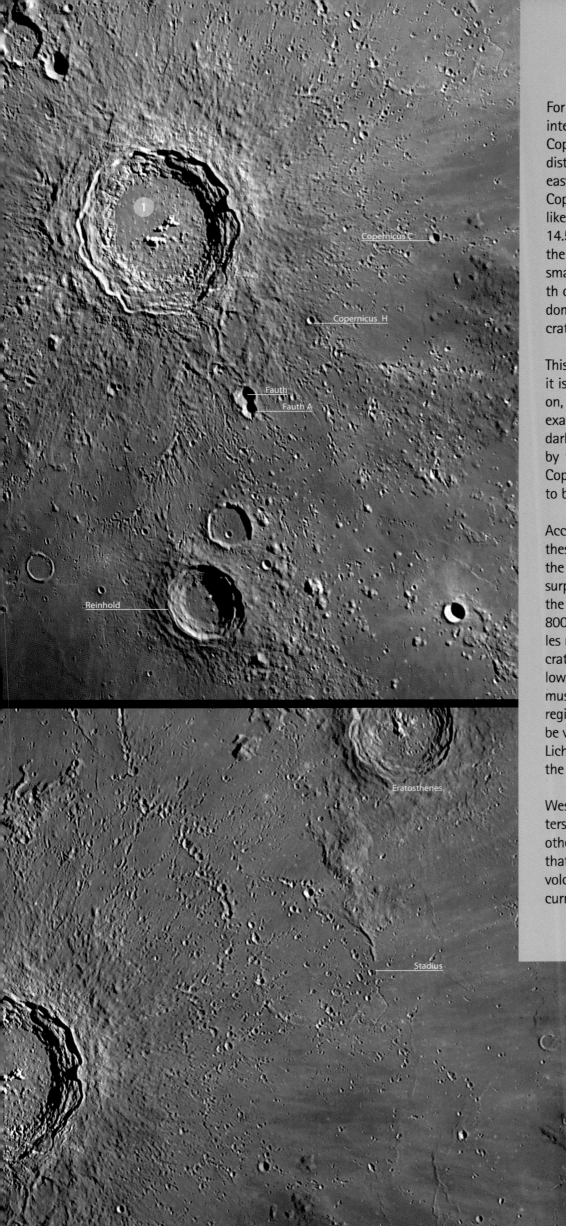

For owners of larger telescopes, a very interesting region lies directly south of Copernicus C (6 km, 7.1°N, 15.4°W). It is distinctive, because of its dark colour, and easy to identify. Immediately to the east of Copernicus C lies a relatively large, dome-like structure with a summit crater (7.0°N, 14.5°W), comparable in its morphology to the Gruithuisen domes, but significantly smaller than the latter. In the region south of Copernicus C lie other smaller lunar domes, of which at least two have summit craters that are about 800 m across.

This region is particularly interesting when it is observed under high solar illumination, and the ray system of Copernicus is examined. While Copernicus C and the dark region to the northeast are overlaid by the Copernicus rays, to the south of Copernicus C the ray system itself appears to be overlaid by the dark ashes.

According to the concept of superposition, these domes must have been active after the Copernicus impact. That is all the more surprising, given that Copernicus is one of the youngest lunar craters, no older than 800 million years, dated from soil samples returned by Apollo 12. In addition, the crater density in the whole region is very low, leading to the conclusion that they must be very young lava flows. Another region of lunar volcanism, also dated to be very young, lies southeast of the crater Lichtenberg on the northwestern limb of the Moon.

West of Copernicus, close to the craters Milichius and Hortensius, there lie other groups of volcanic domes. A theory that would explain the concentration of volcanic structures in Mare Insularum, is currently lacking.

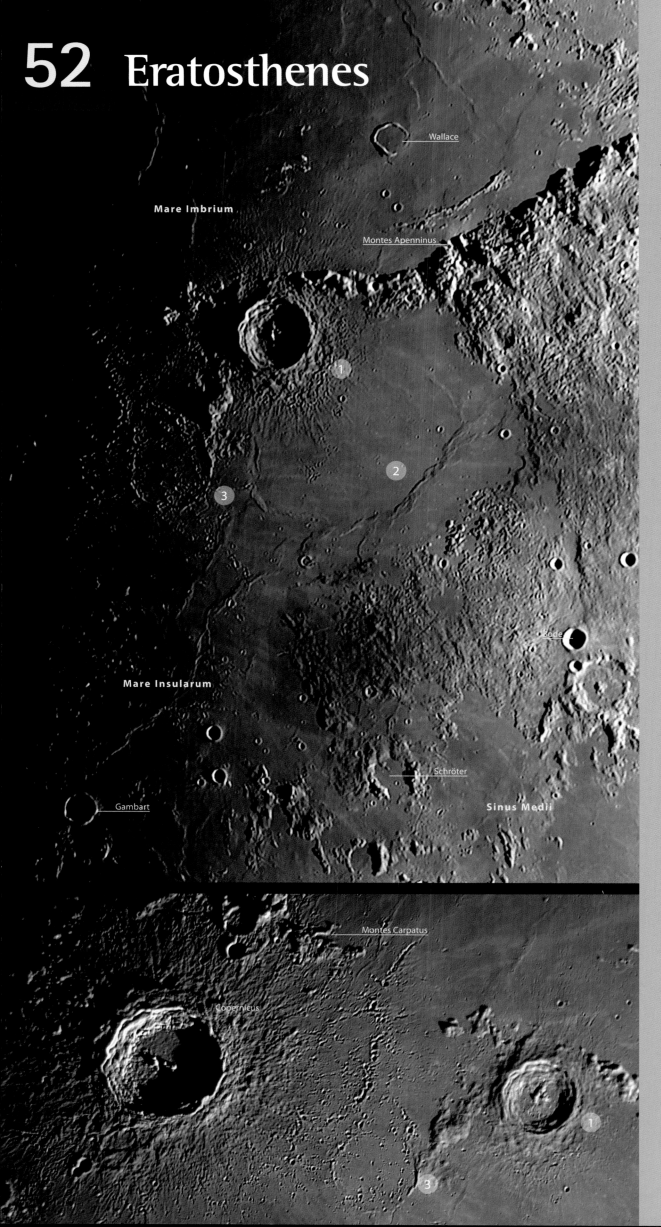

Wallace

Mare Imbrium

Montes Apenninus

1

2

3

Mare Insularum

Bode

Schröter

Gambart

Sinus Medii

Montes Carpatus

Copernicus

1

3

Eratosthenes ①
14.5°N, 11.3°W

Eratosthenes is a typical, little-eroded impact crater with a diameter of 58 km. The depth of the crater floor below the rim is 3.6 km. It has a continuous crater rim and strongly terraced inner walls. The central peak reaches a height of 1.5 km above the crater floor. Eratosthenes was formed about 3.2 billion years ago (within the Eratosthenian period on the lunar time-scale). Because of its greater age – when comparted with Copernicus (800 million ye-ars) – it no longer shows any obvious ray system. The sys-tem that formerly existed has, over the course of billions of years, been darkened by lunar erosion and covered by later impacts.

The foothills of the Apennines reach as far as the northeast-ern wall of the crater. On the southwest lies a nameless ran-ge of mountains which stret-ches south as far as the ghost crater Stadius. The view of Eratosthenes changes great-ly throughout a lunation, and at Full Moon it is practically invisible.

Sinus Aestuum ②
10.9°N, 8.8°W

Sinus Aestuum ('Bay of Bil-lows') is a mare-like plain, about 290 km across. It lies between Eratosthenes and the western slopes of the Apenni-nes. The surface shows a few craters and a few ridges simi-lar to mare wrinkle ridges. The low crater count suggests that the lava flows in Sinus Aestu-um are relatively young.

South of Copernicus C and on the eastern and southeast-ern edges of Sinus Aestuum – bordering on Mare Vaporum – there lie three Dark Mantle Deposit areas.

Stadius 10.5°N, 13.7°W

Stadius is one of the most striking examples of a submerged crater, one of the so-called ghost craters. Stadius is a shallow depression with a broken crater wall, 69 km in diameter. The height of the southeastern wall remnant amounts to only about 600 m. The crater is practically completely covered with ejecta from the Copernicus impact and saturated with secondary craters and crater pits. Stretching towards the northwest is a chain of craterlets, the largest of which are Stadius W, J, T, F, S, E, R, P and Q (all about 5 km in diameter). An L-shaped stucture, consisting of a rille, overlaid by a few crater pits, lies just outside the southern wall. The area between Copernicus and Eratosthenes offers a host of details for any size of telescope. Successful observation of Stadius, the craterlets and crater pits requires a low angle of illumination.

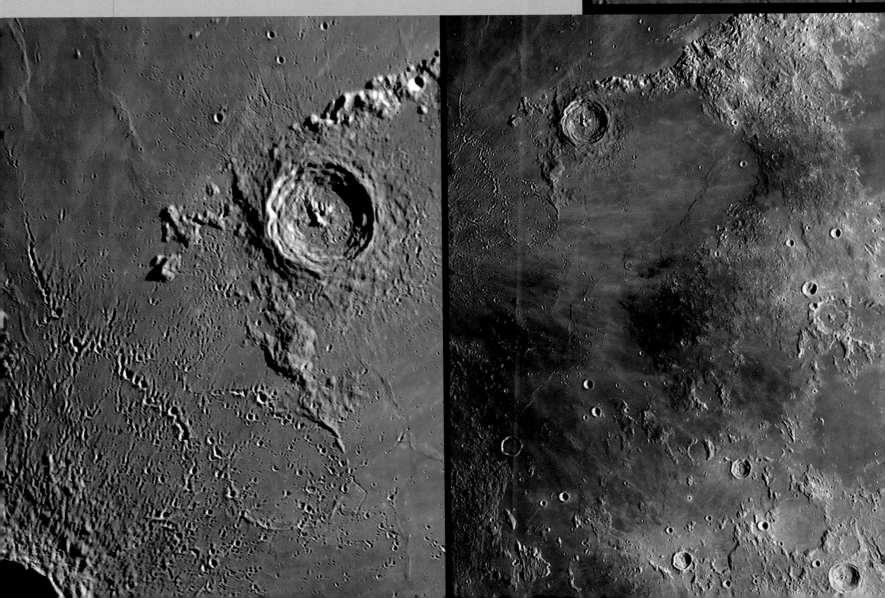

53a Mare Imbrium

Mare Frigoris

Plato

Montes Recti

Mons Pico

Sinus Iridum

Mons Piton

Kirch

Helicon

le Verrier

1

Carlini

Montes Spitzbergen

Aristillus

Heis

McDonald

Autolycus

9

Archimedes

8

10

Montes Archimedes

11

5

Wallace

Montes Apenninus

13

Montes Carpatus

Eratosthenes

Copernicus

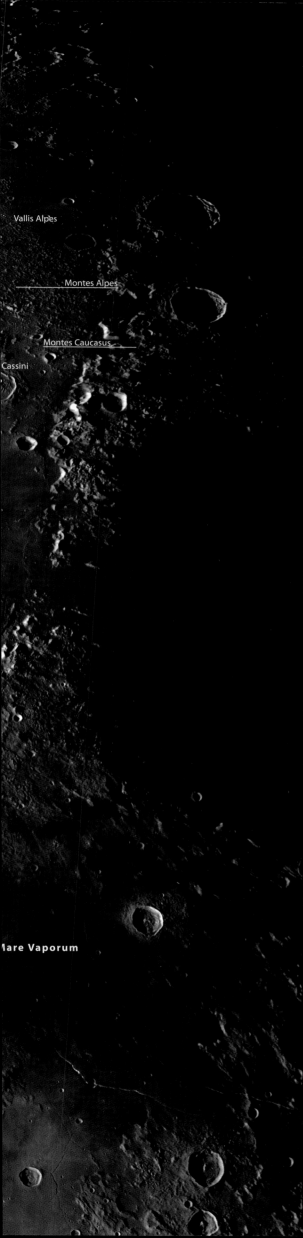

Vallis Alpes

Montes Alpes

Montes Caucasus

Cassini

Mare Vaporum

Mare Imbrium 30.0°N, 20.0°W ①

Mare Imbrium, the 'Sea of Showers' (named by Riccioli) is the dominant feature of the northwestern quadrant of the Moon. With an area of c. 830 000 km², it is – after Oceanus Procellarum – the second largest mare area on the Moon. Like all the other maria it is the lava-filled central portion of a giant impact basin, 1200 km in diameter.

The mountain range of the Alps, Caucasus, Apennines and Carpathians are the remnants of the outer basin ramparts. Remnants of the inner ramparts are the mountain chains Montes Recti, Montes Spitzbergen, Montes Teneriffe and Mons Pico. Towards the west, in the direction of Oceanus Procellarum, there are no visible signs of the basin walls. Many of the isolated mountain massifs, including those along the norther boundary of Mare Imbrium, and Mons Delisle, are the remnants of former highland regions.

The impact that created the basin was so strong that traces of secondary events are still – even after 3.8 billion years – strikingly visible over the whole of the Moon's nearside. These traces have become known as the Imbrium Sculpture. Particularly strikingly visible are the traces of the Imbrium impact in the trio of craters Ptolemaeus, Alphonsus and Arzachel, and in the area around the crater Julius Caesar. The impacting body probably came from a northwesterly direction.

If the Imbrium Basin is observed when the terminator has passed roughly the centre of the Mare, and thus when the Moon's age is about 8.5 days, it is possible to see the boundary between the second of three lava flows that flooded the Imbrium Basin. Southeast of the crater le Verrier (20 km), and between the two small craters Carlini (11.5 km) and McDonald (8 km) a subtle change in the shadows reveals the border of the lava flow. Apollo images show that the origin of this lava lay near the crater Euler. Radar measurements made during the Apollo missions that flew over this area established that the depth of the lava layer amounts to just 35 m. Under low solar illumination this slope, despite its low difference in height, throws a shadow that is many hundreds of metres long, and is thus visible even in amateur telescopes.

To have covered such large areas the lava must have been very fluid – comparable with hot motor oil. The flow velocity must have been correspondingly high: many kilometres per hour. The lava flows lasted for many hundred million years.

Nowadays, multispectral geological maps – for example those from the Clementine mission – enable one to differentiate between the individual lava flows in Mare Imbrium (and not just there) and to date their sequence in time..

Dorsum Zirkel 28.1°N, 23.5°W ②
Dorsum Heim 32.0°N, 29.8°W ③

Two distinctive mare wrinkle ridges in the western portion of Mare Imbrium. Dorsum Zirkel reaches an overall length of about 190 km. Dorsum Heim is about 150 km long. North of the crater C. Herschel lies another, nameless system of wrinkle ridges, which reaches into Sinus Iridum.

Lambert 25.8°N, 21.0°W ④

Lambert is a prominent crater with a diameter of 30 km, terraced inner crater walls and roughly structured ejecta deposited around the crater. South lies the ghost crater Lambert R, with a diameter of 55 km, which is filled with lava up to the crater rim. The Dorsum Zirkel wrinkle ridge lies to the northwest, and a nameless wrinkle ridge is located southeast of the crater.

Euler 23.3°N, 29.2°W ⑤

A crater, 29 km in diameter, with terraced walls, a central peak and a collapsed crater wall. A short chain of craterlets lies southeast of Euler.

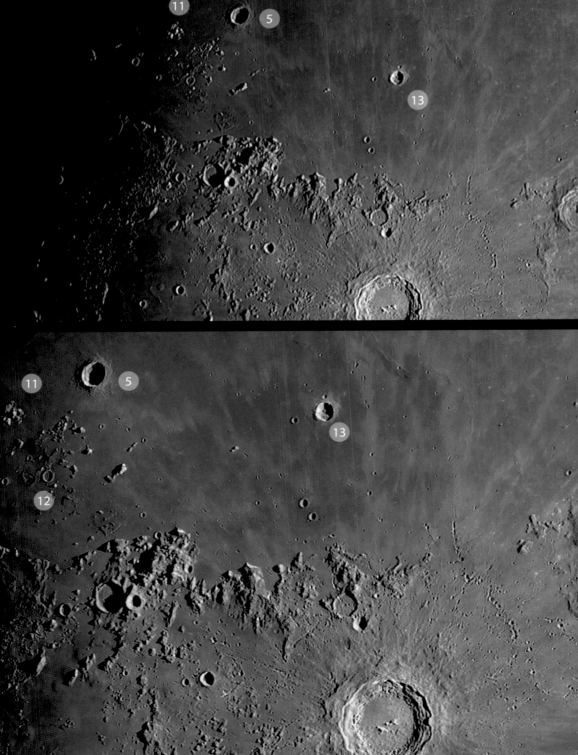

Mons Delisle

Louise

C. Herschel
34.5°N, 31.2°W

A small crater, only 13 km across, named after Caroline Herschel, the sister of William Herschel, who discovered the planet Uranus in 1781. Caroline Herschel was her brother's lifelong assistant and herself discovered eight comets.

Mons la Hire
27.8°N, 25.5°W

An isolated mountain massif occupying an area of about 11 × 25 km. It is probably the remnant of the former highlands (as was Mons Vinogradov) that was largely submerged by the Imbrium lava. The mountain peaks reach a height of 1.5 km above the lava plains.

Diophantus 8
27.6°N, 34.3°W
Delisle 9
29.9°N, 34.6°W

A pair of young craters lying directly north-south of one another, with sharp rims and diameters of 17 and 25 km, respectively. Delisle has an uneven crater floor with a half-destroyed craterlet, lying centrally on the floor.

Mons Delisle is an isolated remnant of the former highlands, lying southwest of Delisle, with a length of about 30 km. The southern portion of the mountain is rounded and higher, towards the north the ridge distinctly flattens out.

Samir (2 km) is a crater pit with a faint ray system under high solar illumination, and directly alongside lies the crater Louise. Diameter measurements for Louise vary between 800 m and 1.5 km. Both craters lie directly in the centre between Diophantus and Delisle.

Two very small sinuous rilles, Rima Diophantus (150 km, 29.0°N, 33.0°W) and Rima Delisle (60 km, 31.0°N, 33.0°W) are visible only in very large telescopes. Their maximum widths occasionally amount to only about 500 m.

Timocharis 26.7°N, 13.1°W ⑩

Timocharis is a young crater, 33 km in diameter, with sharply bounded crater walls. The inner walls are terraced. The floor appears uneven and in the centre a crater pit about 4 km in diameter may be observed. At high solar illumination (Full Moon) bright rays may be observed radiating from Timocharis.

Mons Vinogradov 22.4°N, 32.4°W ⑪

An isolated group of mountains covering an area about 25 km across. The highest peaks reach a height of 1.4 km above the lava plain. The group of mountains is probably a remnant of the former highland region.

Natasha 20.0°N, 31.3°W ⑫

A rather inconspicuous small crater, 12 km in diameter. The crater wall reaches a height of just 300 m, the floor is completely lava-flooded. Large telescopes show a group of crater pits on the crater's floor. West of the crater rim lies Rima Wan-yu (20.0°N, 31.5°W), a rille, only 12 km long, and very difficult to observe. Southeast of Natasha, near the crater T. Mayer, lies a large megadome plateau.

Pytheas 20.5°N, 20.6°W ⑬

A young, 20-km crater with sharp crater walls and a hummocky crater floor. To the east lies a streak of bright ejected material, whose origin is uncertain.

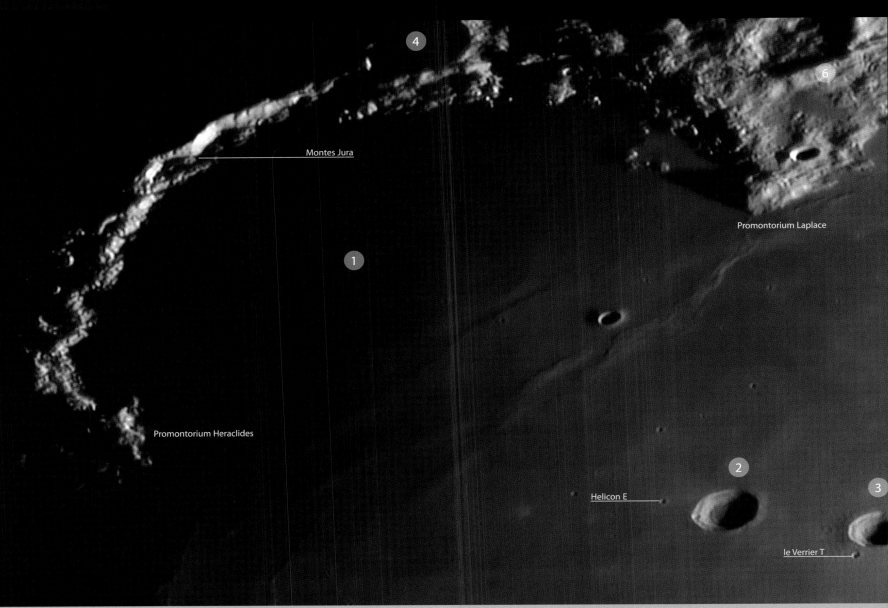

Montes Jura

Promontorium Laplace

Promontorium Heraclides

Helicon E

le Verrier T

Sinus Iridum 45.0°N, 32.0°W

Sinus Iridum, the 'Bay of Rainbows', is the most striking landmark in the northwestern portion of Mare Imbrium. It is the remnant of a giant crater, bordered on the northwest by the Montes Jura (the Jura Mountains), which are undoubtedly a portion of the original crater wall. On the lunar timescale, Sinus Iridum was formed after the Imbrium impact, but before the various lava flows flooded the Imbrium Basin. Sinus Iridum is about 260 kilometres in diameter, and so may be classified one of the smaller impact basins.

The Jura Mountains end in two capes; in the southwest lies Promontorium Heraclides with a height of 1.7 km, and in the southeast lies Promontorium Laplace, which reaches a height of 2.6 km above the average level of the surrounding surface. At sunrise, Promontorium Laplace casts a triangular shadow, and in larger telescopes a lunar dome becomes visible to the west. Other, difficult to observe, domes line near Promontorium Heraclides. They are, however, only observable under extreme grazing illumination. The lava plain within Sinus Iridum is not broken by any large craters. Large telescopes do, however, reveal a series of craterlets and crater pits on the lava surface.

If the northwestern wall of Sinus Iridum is examined more closely, it will be realized that the height of the wall on the eastern side as far as Cape Laplace is approximately the same height, whereas on the west, towards Cape Heraclides it continuously lowers. This effect is possibly the result of a gigantic subsidence that affected the whole southeastern wall and led to Sinus Iridum being flooded by Imbrium lavas from the south.

Under low solar illumination, a series of striking mare wrinkle ridges are visible within Sinus Iridum, which are all largely parallel to one another and to the Jura Mountains. The wrinkle ridges that cross Sinus Iridum to the southeast are related to the inner wall of the Imbrium Basin.

In a narrow window at sunset over Sinus Iridum, just the uppermost crest of the Jura Mountains is illuminated, and is a prominent object. In the 17th century, Cape Heraclides was given the nickname 'the Moon Maiden' by the French astronomer Cassini. Small telescopes under poor seeing conditions showed, with some imagination, a woman's head looking into the bay – although only when viewed with an astronomical (inverted) image orientation.

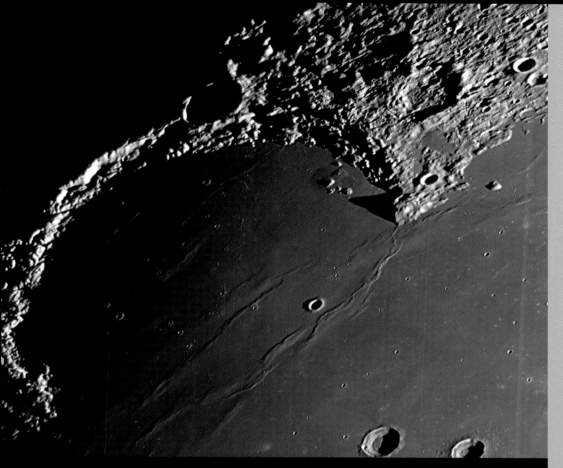

Helicon 40.4°N, 23.1°W ②
Le Verrier 40.3°N, 20.6°W ③

Two proment craters in Mare Imbrium, with similar structures, lying next to one another and orientated directly east-west. Helicon has a diameter of 25 km and a craterlet on the western crater wall. To the west lies Helicon E, with a diameter of only 2.5 km. Le Verrier, with a diameter of 20 km is somewhat the smaller of the two. Directly south lies the craterlet of le Verrier T, with a diameter of about 2 km. Both craters must be comparatively young, because the crater walls are still sharp and clear-cut.

Bianchini 48.7°N, 34.3°W ④
Sharp 45.7°N, 40.2°W ⑤

Two craters, with diameters of 38 km and 39 km, respectively, both lying on the Jura Mountains. Sharp has a central peak, while the crater floor in Bianchini is relatively smooth and level. A small crater chain lies northwest of Sharp.

Maupertuis ⑥
49.6°N, 27.3°W
La Condamine ⑦
53.4°N, 28.2°W

Maupertuis (46 km) and La Condamine (37 km) are older than Sinus Iridum, because they have been severely eroded by ejecta from the Imbrium impact. Maupertuis has been almost completely destroyed, and is of a vaguely pentagonal shape, lying near Promontorium Laplace. A small, smooth lava surface lies directly south of the crater. A narrow rille system (100 km, 51.0°N, 22.0°W) runs along the southeast side.

La Condamine has no central peak, but which must have existed, given the size of the crater. High-resolution images from the Lunar Orbiter probes do, however, show a ring of hills and rilles on the crater's floor. The crater was altered by volcanism after its formation and is included in the category of FFC craters.

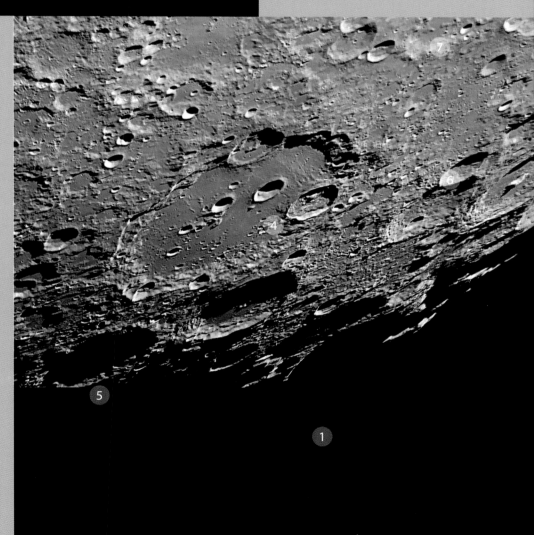

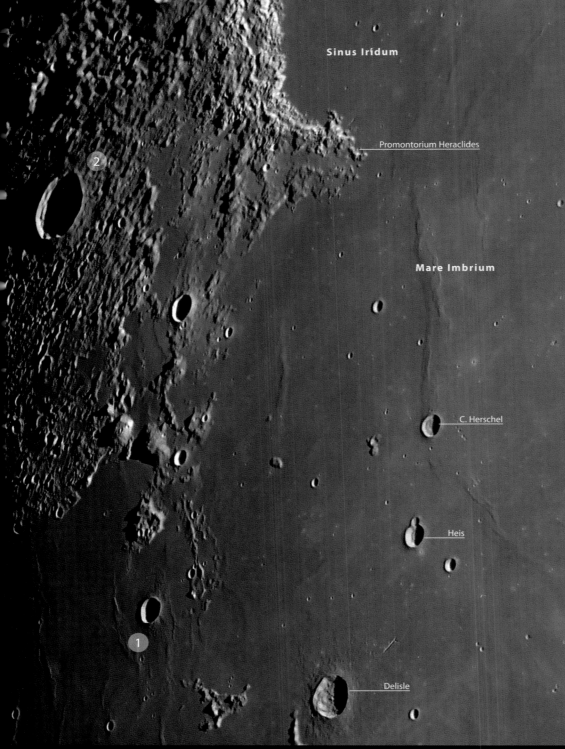

Sinus Iridum

Promontorium Heraclides

Mare Imbrium

C. Herschel

Heis

Delisle

Gruithuisen 32.9°N, 39.7°W ①

Gruituisen is a small crater of 16 km in diameter and a depth of about 1.9 km, lying at the northern end of Dorsum Bucher. It was probably formed during the Eratosthenian period on the lunar timescale.

The crater was named after Franz von Paula Gruithuisen (1774–1852), professor of astronomy and lunar observer, who, however, after publications about the 'Discovery of many clear traces of inhabitants on the Moon and a colossal artificial structure of theirs' exposed himself to ridicule, because it had been known for a long time that there is neither air nor water on the Moon.

The crater Gruithuisen is a classic example of a simple crater, according to the USGS standard sequence for lunar craters.

Mairan 41.6°N, 43.4°W ②

Mairan is a relatively young crater, 40 km in diameter, with a sharply defined crater wall and a slightly eccentrically placed, small central peak. Mairan T, with a diameter of only 3 km is a volcanic dome with a summit crater, lying west of Mairan. The rille Rima Mairan (38.0°N, 47.0°W) lies south of Mairan T. It is a linear rille, with a maximum width of 2 km, that runs for an overall length of 90 km. Directly northwest of Mairan lies the almost completely destroyed crater Louville, with a diameter of about 36 km and the two craterlets Louville A and B (both about 8 km across). Louville is difficult to locate in its extemely craggy surroundings.

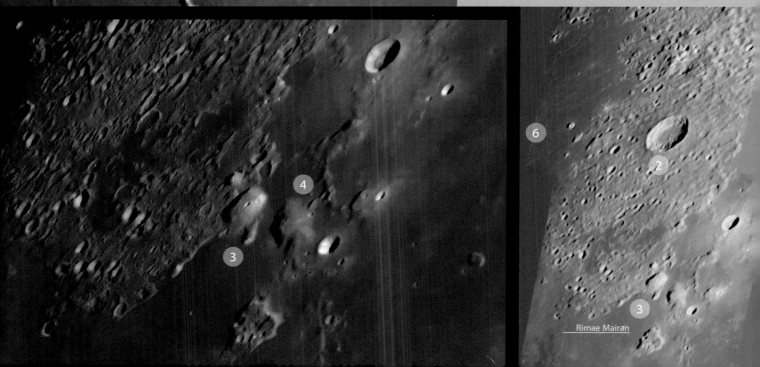

Rimae Mairan

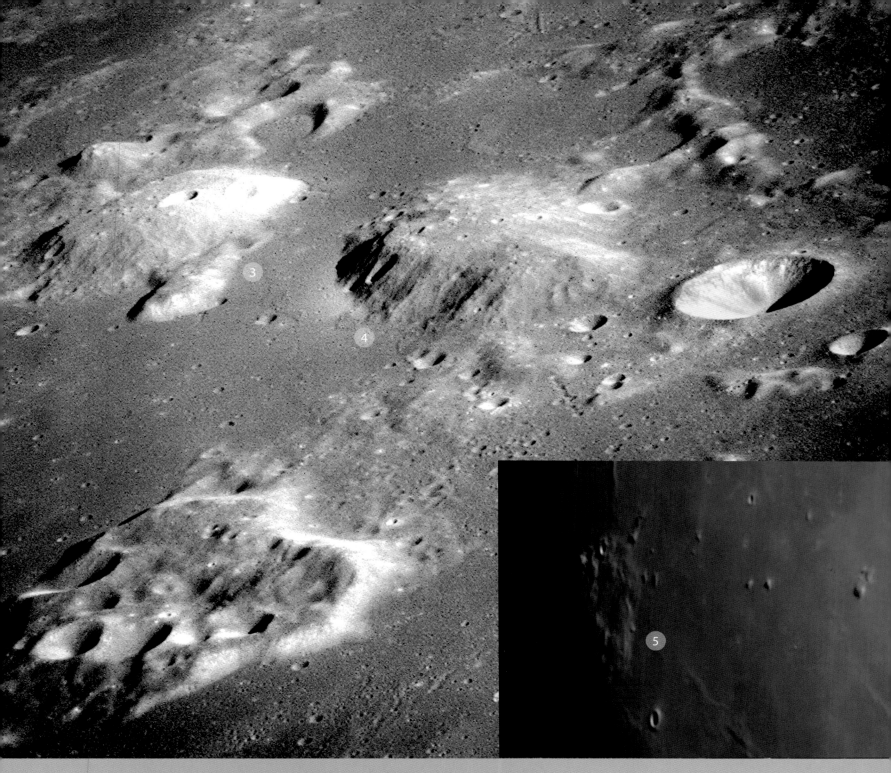

Mons Gruithuisen Gamma 36.6°N, 40.5°W ③

A volcanic dome (an extrusion of magma) with a base of approximately 20 × 20 km, lying between the craters Gruithuisen and Mairan. Its height amounts to about 1 km. Because of limb-foreshortening, its shape appears somewhat elliptical and resembles an upturned bathtub. The ridge has a small summit crater, about 1 km in diameter, and a small hollow, that is slightly smaller, and is naturally difficult to observe. While the mountain itself is visible in even the smallest telescopes, the summit crater and hollow qualify as a test of resolution for a good 250-mm telescope.

Mons Gruithuisen Delta 36.0°N, 39.5°W ④

A lunar dome, similar to Gruithuisen Gamma, although irregular in shape. The diameter of the base amounts to about 20 km.

Mons Rümker 40.8°N, 58.1°W ⑤

Mons Rümker is a unique feature on the Moon. It is the prototype of a volcanic megadome, and has a base diameter of about 70 km,

with a height of only 700 m. Because of its limited height and its position near the limb it is a difficult object to observe.

Mons Rümker lies on the boundary between Sinus Roris and Oceanus Procellarum. The surface of the dome shows a slight depressed area, many summit craters and boulders. Under very low illumination (at sunrise or sunset) a complex system of mare wrinkle ridges becomes visible. Favourable libration conditions are required for observation.

Rima Sharp 46.7°N, 50.5°W ⑥

A sinuous, very narrow rille, beginning west of Mairan (between Mairan T and Mairan G) that stretches, with meanders, northwards for an overall length of 110 km. It ends at the small crater Louville D (7 km). Rima Sharp is 500 m wide at maximum, and requires a very large telescope aperture (at least 400 mm) to be observed.

Mare Frigoris 56.0°N, 1.4°E

Whereas the other maria show a tendency towards a nearly circular shape, Mare Frigoris ('Sea of Cold') is elongated and lies in the Moon's northern polar region. It stretches about 300 km in a north–south direction, but over 1000 km in an east–west direction. The areas of Lacus Mortis and Sinus Roris form part of it. It covers an area of about 440 000 km² and is approximately comparable with the area of the Black Sea on Earth.

J. Herschel 62.0°N, 42.0°W

The remnant of a giant crater, 165 km in diameter. The crater floor is covered in rubble from the Imbrium impact. In larger telescopes and under grazing illumination the rough nature of the rubble is clearly visible.

There are three craters on the Moon's nearside that are named after members of the Herschel family. J. (John) was the son of William Herschel (the crater Herschel north of Ptolemaeus) and C. (Caroline) Herschel, his sister (the crater in Mare Imbrium).

Harpalus 52.6°N, 43.4°W

A prominent, young crater with a sharp crater wall, 39 km in diameter, lying on the lava surface of Mare Frigoris. The crater's floor appears rough and furrowed. Under high illumination the crater exhibits a ray system. To the west lie a few smaller, low, isolated ridges.

Horrebow 58.7°N, 40.8°W

A small crater, 24 km in diameter, which overlays Horrebow A (25 km) to the northeast. Both craters lie on the southern crater wall of J. Herschel. Together, Horrebow and Horrebow A appear rather pear-shaped, like the crater Torricelli.

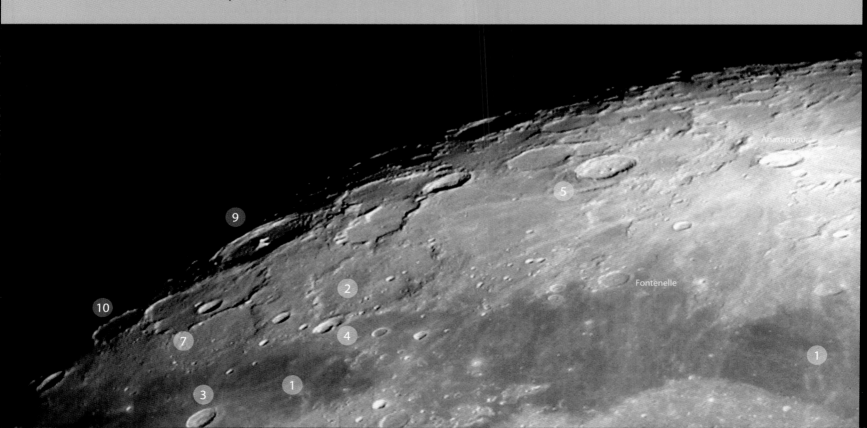

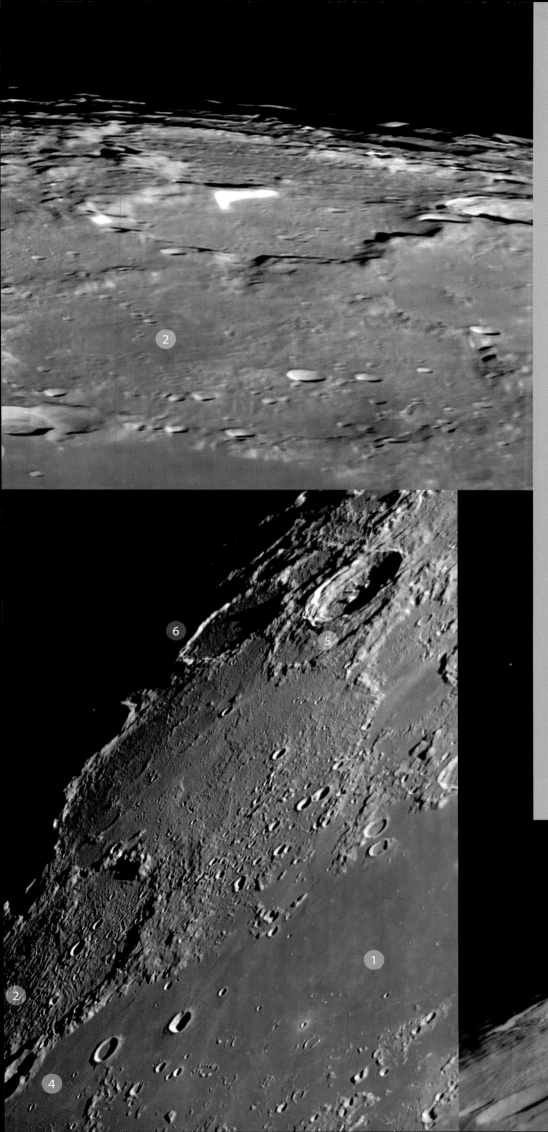

Philolaus 72.1°N, 32.4°W ⑤
A crater, 70 km in diameter, with terraced inner crater walls and two central peaks on the crater's floor.

Anaximenes 72.5°N, 44.5°W ⑥
A crater with a diameter of 80 km. The walls of the crater are almost completely eroded, the crater's floor is flat and level.

Babbage 59.7°N, 57.1°W ⑦
A large complex crater, 143 km in diamater. Lying on the crater's floor are the circular craters Babbage A (32 km) and Babbage C, a craterlet 14 km in diameter.

South 58.0°N, 50.8°W ⑧
Like Babbage, South is the remnant of a large complex crater, 105 km in diameter, bordering on Babbage to the northwest. A small crater chain lies at the edge of the southwestern crater wall.

Pythagoras 63.5°N, 63.0°W ⑨
A large, and despite its position near the limb, a prominent complex crater, diameter 142 km. Two pyramid-shaped, massive cental peaks lie next to one another on the crater's floor.

Oenopides 57.0°N, 64.1°W ⑩
A complex crater, 67 km diameter. The southwestern crater wall has been almost completely destroyed by overlapping craterlets. Southwest of Oenopides there are also groups of craterlets and crater pits.

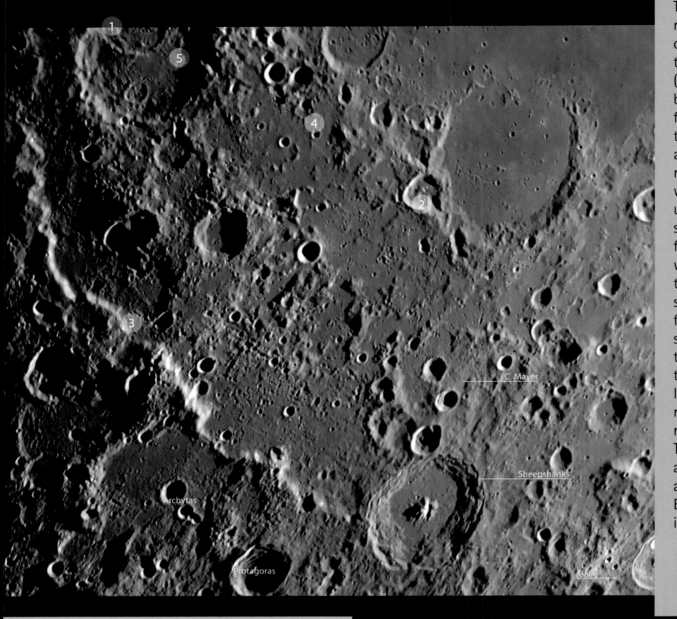

North Pole 90°N ①

The Moon's North Pole lies north of Mare Frigoris in an old highland region. The craters in this area are often flat (lava-flooded), heavily eroded by secondary impacts and frequently overlap one another. Some, such as W. Bond and J. Herschel, have very rough crater floors, heaped with ejecta from the Imbrium impact, whereas others, such as Meton, have smooth floors and have been covered with Imbrium lava. The whole terrain and many crater rims show structures with valleys, fissures and crater chains, similar to the structures in the southern Imbrium Sculpture near J. Caesar and Ptolemaeus. Because of limb foreshortening the traces in the north are harder to observe. The most clearly recognized are valley-like structures east and west of the crater W. Bond and a long crater chain in J. Herschel.

Meton 73.6°N, 18.8°E ②

Meton is the heavily eroded remnant of a very large complex crater, 130 km in diameter. Overlapping it are two other large craters, Meton C (77 km) and Meton D (78 km). All three crater floors appear flat and level. Under low solar illumination and in larger telescopes various craterlets become visible on the crater floors. The trio of craters lies north of Mare Frigoris in the Moon's north polar region. Under high illumination, rays of ejecta from the crater Goldschmidt are visible on the crater floor, and which resemble those in the crater Stöfler in the southern highlands.

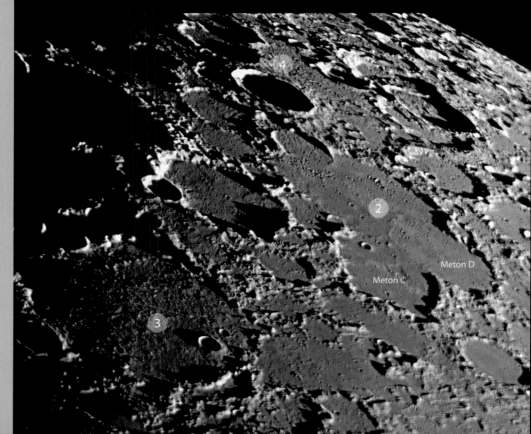

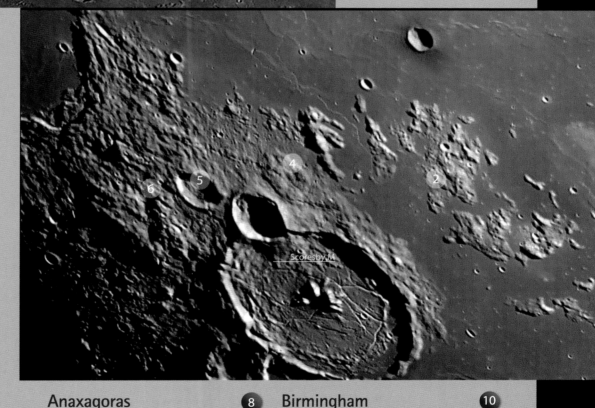

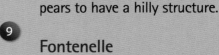

W. Bond 3
65.4°N, 4.5°E

A giant complex crater, 156 km in diameter, lying southwest of Meton. The southwestern crater wall has been broken by the crater Timaeus (32 km). Large portions of the crater's floor are covered with long heaps of ejecta from the Imbrium impact. In large telescopes, a small, nameless rille may be seen on the crater floor.

Scoresby 77.7°N, 14.1°E 4
Gioja 83.3°N, 2.0°E 5
Byrd 85.3°N, 9.8°E 6
Peary 88.6°N, 33.0°E 7

North of Meton these four craters lie in the direction of the North Pole, which is located near Peary. Scoresby is a crater 55 km across, with two low hills and a small crater on the floor. Scoresby M (54 km) lies nearby to the southwest.

Gioja measures 41 km across and shows a heavily eroded western crater wall. A large crater, 93 km in diameter, directly next to Gioja is Byrd. The crater's floor is saturated with crater pits. The eastern and western walls are heavily eroded.

Peary is a 73-km crater, which lies nearly at the Moon's North Pole. Observation from Earth succeeds only when the libration conditions are particularly favourable.

Peary is, in fact, nearly circular, but because of the extreme limb foreshortening it appears distorted into an extreme ellipse. A large part of the southwestern boundary is open to a nameless area of the Moon. Portions of the southern crater floor lie in permanent darkness. Grazing sunlight illuminates two sections of the crater's rim throughout the whole of the lunar summer.

Anaxagoras 8
73.4°N, 10.1°W

A crater with a diameter of 50 km. Under high solar illumination a ray system is visible. One of the rays may be followed as far as the crater Plato.

Goldschmidt 9
73.2°N, 3.8°W

A large crater, 113 km in diameter, broken by the crater Anaxagoras on Goldschmidt's western side. The outer crater wall appears rough and furrowed.

Birmingham 10
65.1°N, 10.5°W

The remnants of a large and heavily eroded crater, 92 km in diameter. Under grazing illumination (when near the terminator) the crater's floor appears to have a hilly structure.

Fontenelle 11
63.4°N, 18.9°W

A 38-km crater with jagged crater walls and with a craterlet in the centre of the crater's floor. It appears significantly elliptical because of the effect of limb curvature on the perspective.

Mare Imbrium

Oceanus Procellarum

Seleucus

Aristarchus 23.7°N, 47.4°W ①

The crater Aristarchus lies on a rhombus-shaped, high plateau in the northern portion of Oceanus Procellarum and is the boundary between the latter and the lava plain of Sinus Roris. Aristarchus is a very young crater (created about 500 million years ago), with a ray system and a diameter of about 40 km. The crater has a small central peak. The crater's floor lies 3 km below the rim, and is deep enough to have exposed the bright anorthositic rocks of the upper lunar crust through the impact and to transport them to the surface. They also form the ray system of ejecta around the crater. The inner crater walls of Aristarchus have a high albedo and are crossed by dark, radial stripes. The crater and its surroundings are visible even under the ashen-coloured light of Earthshine.

The whole plateau was repeatedly the centre of observation of the so-called Lunar Transient Phenomena, luminous events, perhaps caused by the release of radioactive gases from the Moon's interior. During overflights of the plateau by Apollo 15 and the Lunar Prospector probe, measurements of higher traces of radioactive radon gas were detected. On contrast-enhanced images this area is one of the most colourful areas on the nearside of the Moon.

Vallis Schröteri 26.2°N, 50.8°W ②

Schröter's Valley is the largest and longest sinuous rille on the Moon, but its origins are not comparable with valleys on Earth. Vallis Schröteri begins about 30 km north of the crater Herodotus at a small crater, only 6 km across, known as the 'Cobra Head'. The rille is 10 kilometres wide there, and runs fairly straight towards the north, then it bends (directly south of which there is a lunar dome with a summit crater), and turning almost 180°, runs southwards again. The valley ends, at a length of almost 170 km, at a steep escarpment 1 km high on the edge of the plateau. Here the rille is only 500 m wide.

The floor of Vallis Schröteri is extremely smooth and level, and at its head is about 1 km deep. Only towards the end does it become shallower. Another rille runs within the floor of Vallis Schröteri, following the course of the valley.

Schröter's Valley strongly resembles a meandering, dry, terrestrial watercourse. Like Hadley Rille, Vallis Schröteri is also a lava tube, where, after the lava flow has come to an end, the roof has collapsed. The vent that provided the lava was probably the crater at the Cobra Head.

Montes Harbinger 27.0°N, 41.0°W ③

Montes Harbinger are a small group of isolated mountains, extending for about 90 × 40 km, trending in a north-south direction. The peak heights reach 2.5 km. Lunar geologists believe that the Harbinger region is a megadome, an uplifted section of the lunar surface, raised by upwelling masses of lava from the Moon's interior. This theory is unconfirmed and because of the low height of the area, its observation is difficult. The mountain peaks illuminated by the rising Sun announce dawn over the crater Aristarchus.

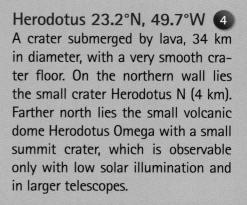

Herodotus 23.2°N, 49.7°W ④

A crater submerged by lava, 34 km in diameter, with a very smooth crater floor. On the northern wall lies the small crater Herodotus N (4 km). Farther north lies the small volcanic dome Herodotus Omega with a small summit crater, which is observable only with low solar illumination and in larger telescopes.

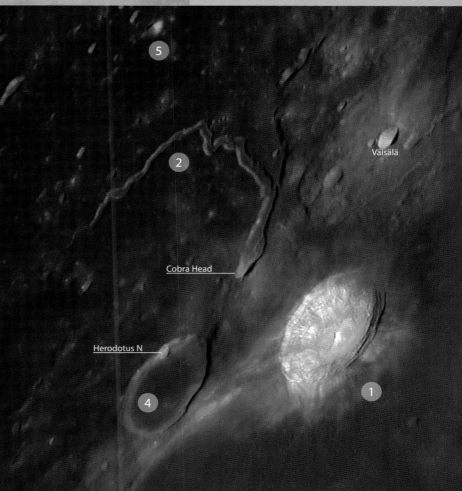

Väisälä

Cobra Head

Herodotus N

Mons Herodotus 27.0°N, 53.0°W ⑤

A small, isolated mountain with a base diameter of about 5 × 5 km.

Montes Agricola 29.1°N, 54.2°W ⑥

A fairly straight, elongated, but very narrow chain of mountains with a length of over 160 km. It lies amid the plains of Oceanus Procellarum. In the northern portion, the chain of mountains runs – at a right angle – through the mare wrinkle ridge Dorsum Niggli (29.0°N, 52.0°W), which runs 50 km to its junction with the Aristarchus plateau.

Rimae Aristarchus 26.9°N, 47.5°W ⑦

A highly branched system of sinuous rilles. They lie between Rupes Toscanelli and Rimae Prinz, and run northwest for a length of 120 km. Here again many of the sections of the rilles are associated by crater pits.

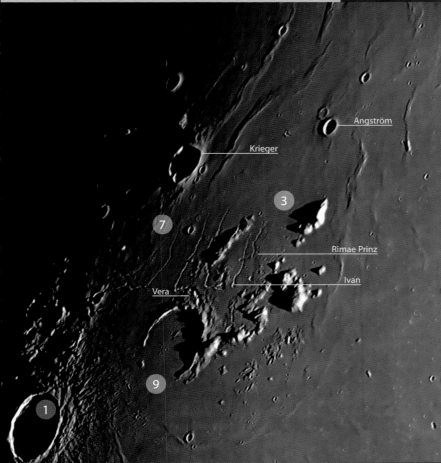

Ångström

Krieger

Rimae Prinz

Ivan

Vera

Rupes Toscanelli 27.4°N, 47.5°W ⑧

An escarpment with a length of about 70 km, bordering the small crater Toscanelli (7 km) to the south.

Prinz 25.5°N, 44.1°W ⑨

Prinz is the horseshoe-shaped remnant of a generally lava-flooded crater, 46 km in diameter. A complex, sinuous rille system, Rimae Prinz, winds over 80 km northwards, starting from the crater wall. The main rille begins at the crater pit Vera (5 km, formerly known as Prinz A). A second section of rille begins at the crater pit Ivan (4 km, formerly Prinz B), also north of Prinz. Observation of Rimae Prinz requires a large instrument. Based on their structure, Vallis Schröteri and Rimae Prinz are very similar, but the rilles of Rimae Prinz are significantly smaller and narrower.

Mare Imbrium

Aristarchus

Reiner Gamma

Oceanus Procellarum

Copernicus

Grimaldi

Mare Insularum

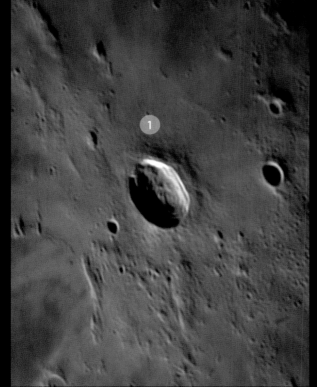

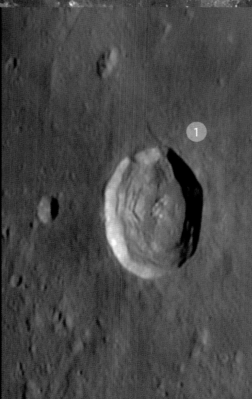

Kepler
8.1°N, 38.0°W

Kepler is undoubtedly one of the most prominent craters in the Oceanus Procellarum region. It is a relatively small crater, 31 km in diameter, with terraced inner walls, central peaks, and an uneven crater floor. Kepler is the centre of a very conspicuous ray system and is therefore one of the youngest craters. The wall of the craters appears to be slightly hexagonal. About 40 km to the northwest, a volcanic dome is visible under low solar illumination and in larger telescopes. It has a small, eccentrically placed summit crater.

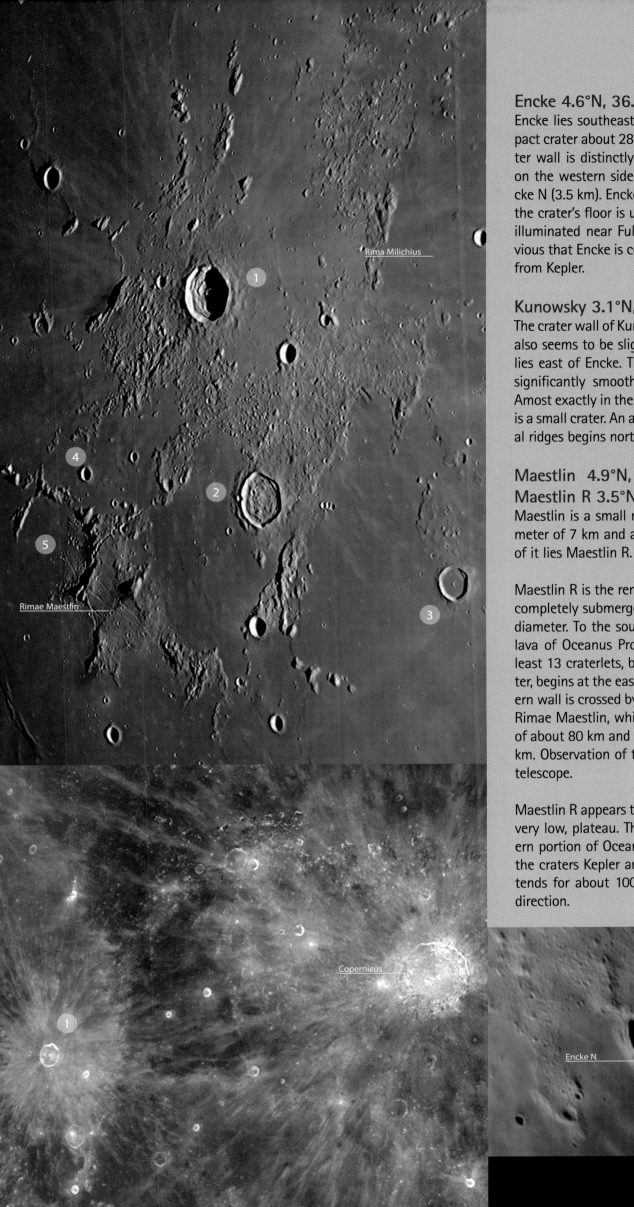

Rima Milichius

Rimae Maestlin

Copernicus

Encke N

Encke 4.6°N, 36.6°W (2)

Encke lies southeast of Kepler and is an impact crater about 28 km in diameter. The crater wall is distinctly six-sided and is broken on the western side by the small crater Encke N (3.5 km). Encke has a central peak, and the crater's floor is uneven and rough. When illuminated near Full Moon, it becomes obvious that Encke is covered by the ray system from Kepler.

Kunowsky 3.1°N, 32.5°W (3)

The crater wall of Kunowsky (18 km diameter) also seems to be slightly six-sided. Kunowsky lies east of Encke. The crater's floor appears significantly smoother than that of Encke. Amost exactly in the middle of the floor there is a small crater. An arcuate chain of individual ridges begins north of the crater rim.

Maestlin 4.9°N, 40.6°W (4)
Maestlin R 3.5°N, 41.5°W (5)

Maestlin is a small round crater, with a diameter of 7 km and a depth of 1.5 km. South of it lies Maestlin R.

Maestlin R is the remnant of a crater, almost completely submerged by lava, and 61 km in diameter. To the southwest it is open to the lava of Oceanus Procellarum. A chain of at least 13 craterlets, broken by one larger crater, begins at the eastern wall. The southeastern wall is crossed by a system of linear rilles, Rimae Maestlin, which has an overall length of about 80 km and a maximum width of 1.5 km. Observation of the rilles requires a large telescope.

Maestlin R appears to lie on an extensive, but very low, plateau. This lies in the southeastern portion of Oceanus Procellarum, west of the craters Kepler and Encke, and which extends for about 100 km in the north-south direction.

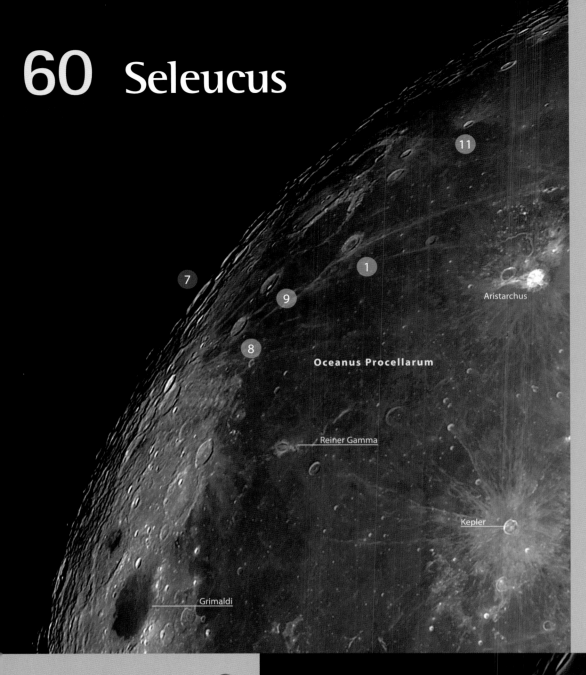

Oceanus Procellarum

Aristarchus

Reiner Gamma

Kepler

Grimaldi

Seleucus 21.0°N, 66.6°W ①

Seleucus is a prominent crater, east of Eddington, with a diameter of 43 km. It exhibits terraced inner walls and a small central peak. The eastern wall is covered in bright streaks, consisting of ejecta that originated in the distant crater Glushko (formerly known as Olbers A, 43 km).

Briggs 26.5°N, 69.1°W ②

A crater, 37 km in diameter, with a mountain ridge on its floor.

Eddington 21.3°N, 72.2°W ③
Struve 22.4°N, 77.1°W ④
Russell 26.5°N, 75.4°W ⑤

Eddington, Struve and Russell are impressive examples of large, lava-flooded craters with completely smooth floors. They lie close to the northwestern limb of the Moon, and are best observed under favourable libration angles. The crater diameters are 118, 164 and 103 km respectively. Struve and Russell merge into one another, and Eddington lies to the east of Struve. The 80° lunar meridian runs along the western wall of Struve.

Schiaparelli ⑥
23.4°N, 58.8°W

Schiaparelli is a conspicuous crater, 24 km in diameter, named after the Italian astronomer Schiaparelli, who initiated the long debate over martian canals. Schiaparelli lies at the end (or the beginning) of the 200-km long mare wrinkle ridges known as Dorsa Burnet. A bright streak of ejected material ends southwest of Schiaparelli. To the east lies the prominent Aristarchus plateau with Vallis Schröteri.

Vasco da Gama ⑦
13.6°N, 83.9°W

A crater lying extremely near the limb, 83 km in diameter, and with a central peak.

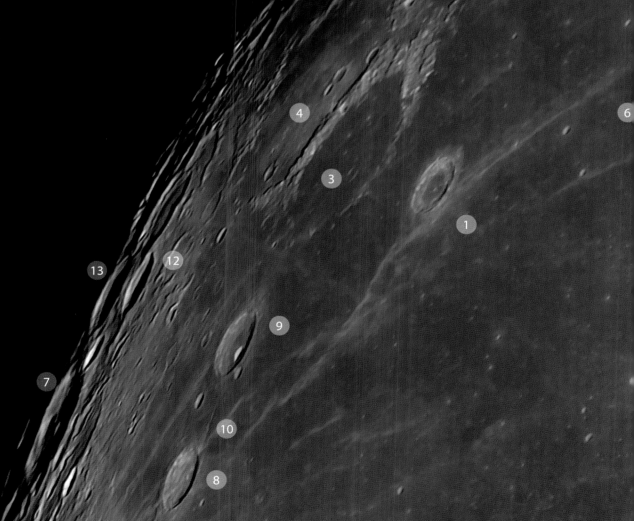

Cardanus 13.2°N, 72.5°W (8)

A crater, 49 km in diameter, with sharp-edged crater walls and a small central peak. Northeast of the crater's rim lies a ghost crater, completely submerged by lava. Lunar Orbiter images show that the crater's floor is crossed by a rille system, and so Cardanus must be classified as a FFC crater.

Krafft 16.6°N, 72.6°W (9)

A crater, 51 km in diameter, with sharp-edged crater walls. Krafft C (13 km), with an equally sharp rim, lies on the crater's floor.

Catena Krafft 15.0°N, 72.0°W (10)

Catena Krafft is a 60-km long crater chain consisting of heavily eroded, partially overlapping craterlets. Running along the 72° meridian, they join the craters Krafft and Cardanus. The diameter of the craters reduces from south to north. The chain of craters runs onto the floor of Krafft, and cuts through the crater Krafft C. Under poor seeing conditions, or with smaller telescope apertures, Catena Krafft appears like a broad linear rille.

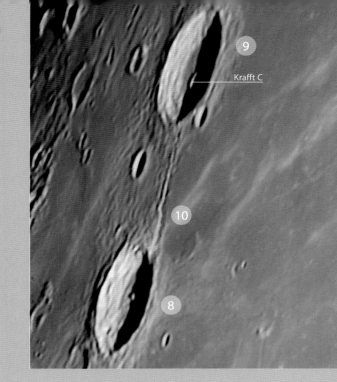

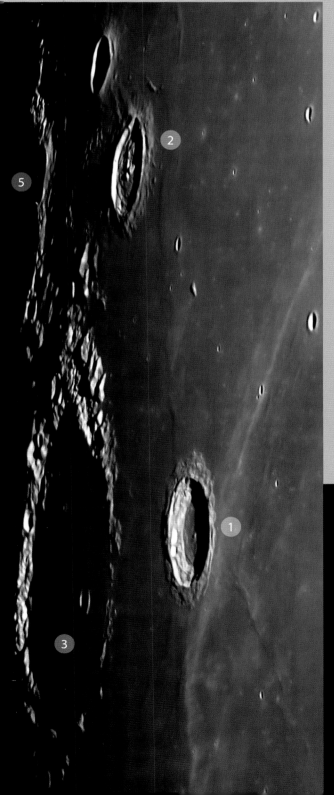

Lichtenberg 31.8°N, 67.6°W (11)

Lichtenberg is an interesting crater, 20 km in diameter, in Oceanus Procellarum at the border with Sinus Roris. It is a young crater, because it exhibits a bright ray system, and its formation has been dated to the Copernican period on the lunar timescale. What is interesting is that the ray system is prominent to the north, west, and south, but is non-existent on the southeast. There, the bright ejected material is covered by dark lava, that emerged after the formation of the crater. This fact indicates that there were still major lava eruptions long after the large basins were flooded, 3.8 to 2.5 billion years ago. The Lichtenberg lava is the youngest and has been dated to an age of about 900 million years.

Dalton 17.1°N, 84.3°W (12)
Einstein 16.3°N, 88.7°W (13)

Dalton and Einstein lie extremely near the limb and Eistein in particular may be observed only under the most favourable libration conditions. Dalton is a crater 60 km across, and is overlaid by Einstein on its western side. Because of the effect of limb curvature on the perspective, Einstein appears exceptionally elliptical, but, as images from lunar probes have shown, it is a large circular crater of about 200 km in diameter. In the centre of the crater's floor, lies Einstein A, with a diameter of 51 km and a central peak.

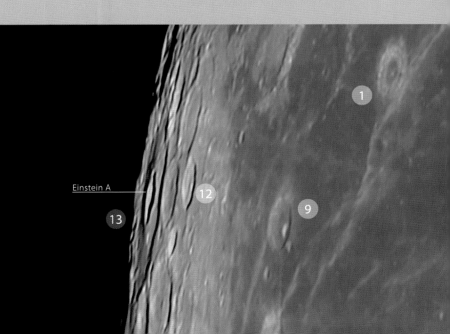

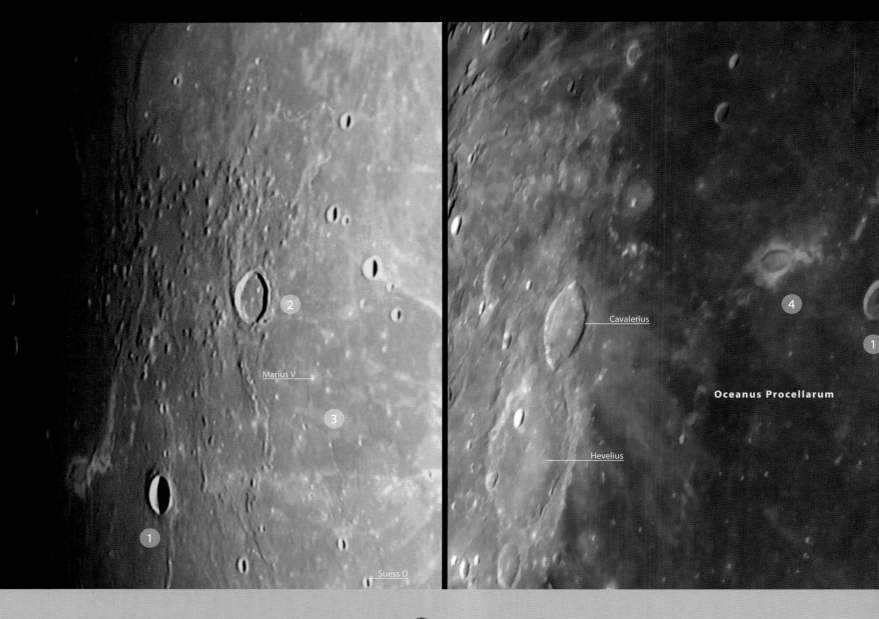

Reiner 7.0°N, 54.9°W 1

A crater, lying isolated in Oceanus Procellarum, 29 km in diameter. It is circular, but because of limb foreshortening appears significantly elliptical. The crater's floor is fairly rough and has – lying centrally – an elongated central mountain, orientated north-south.

Marius 11.9°N, 50.8°W 2

The crater Marius has a diameter of 41 km, a flat, smooth floor and a very low wall height above the crater's floor (1.6 km). Marius G (about 3.3 km) lies on the northern floor and the southeastern wall is breached by a small, slightly elliptical crater Marius H (5 km). Farther west of Marius lie the smaller craters Marius A and Kepler C (both about 10 km). They are linked by a ray of bright ejected material, part of Kepler's ray system.

The area south of Marius is rich in mare wrinkle ridges. North and west of the crater lies the largest group of related lunar domes. In total, Apollo images show over 300 of these volca-

nic structures, which are all only 200 to a maximum of 300 m high. The best observational conditions are about 12 to 13 days after New Moon, when the Sun illuminates the area at a grazing angle.

A complex, nameless system of mare wrinkle ridges begins south of Marius, and at 9.0°N, 51.8°W there lies a larger lunar dome with a summit crater.

Rima Suess 6.7°N, 48.2°W 3

A very narrow rille, whose observation requires a telescope with an aperture that is significantly larger than 250 mm. It lies south of Marius and begins near the craterlet Marius V (2 km, bright halo, near the impact site of Luna 7) and runs in a meandering fashion for a length of about 200 km in a southerly direction. It ends near the crater Suess D (c. 9 km). The crater Suess (12 km) is located about 50 km west of Suess D, and in between there is a section of mare wrinkle ridge.

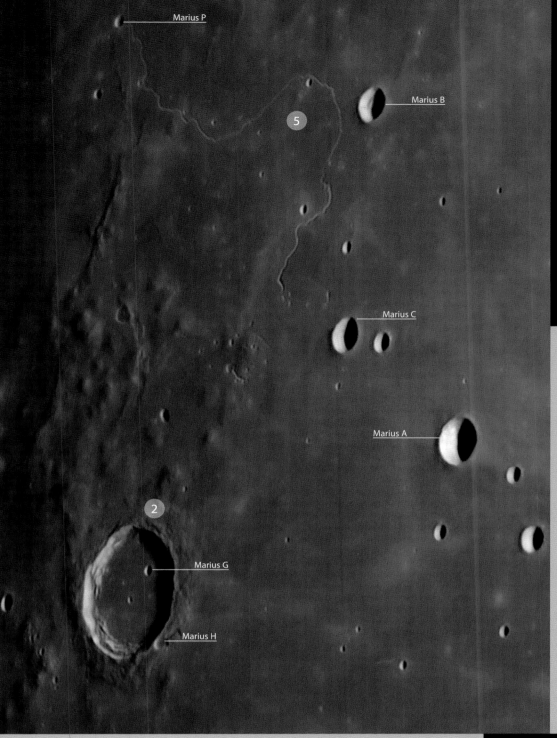

Marius P

Marius B

5

Marius C

Marius A

2

Marius G

Marius H

Rima Marius 5
17.0°N, 49.0°W

The Marius Rille is one of the most prominent examples of a sinuous rille. It begins at the crater Marius C (11 km), about 25 km north of Marius, and where its width is about 2 km. It initially runs northwards, before it curves west at the crater Marius B (12 km). Here it narrows to just about 1 km in width, and runs directly towards the craterlet Marius P (4 km). Here it again turns west and ends, with a width of about 500 m about 40 km west of Marius P. Its overall length is about 250 km. The beginning of the rille is visible with medium-sized telescopes, but the end requires large apertures and good seeing conditions.

Reiner Gamma 7.5°N, 59.0°W 4

Reiner Gamma is one of the most unusual features on the nearside of the Moon, and is observable even with small telescopes. It is a so-called swirl.

Reiner Gamma consists of a layer of very bright material and shows no traces of any form of relief, neither on detailed images of this region from the Lunar Orbiter probes, nor on those taken during the Apollo missions. The difference in height from that of the surrounding terrain is essentially zero. The central portion of the feature is a bright ellipse, with a 70-km long streak running towards the north. Reiner Gamma lies west of the crater Reiner.

Reiner Gamma is the centre of a magnetic anomaly (a magcon). Whereas the normal lunar magnetic field at the surface is nearly zero, the magnetic field strength increases significantly over this object. The origin of Reiner Gamma remains unexplained to this day. Theories range from a cometary impact to shock waves released by an impact on the Moon's farside. Another magnetic anomaly is exhibited by the bright spot on the wall of the crater Descartes.

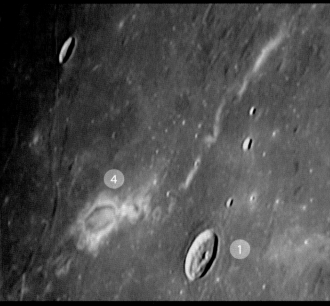

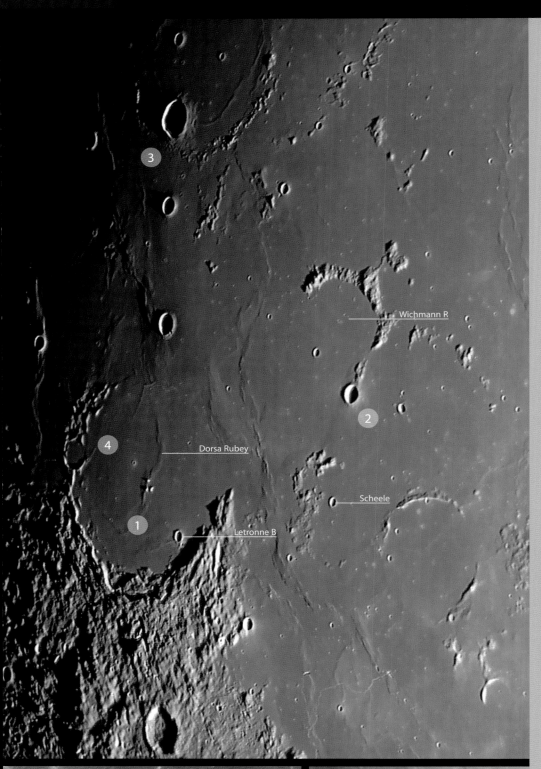

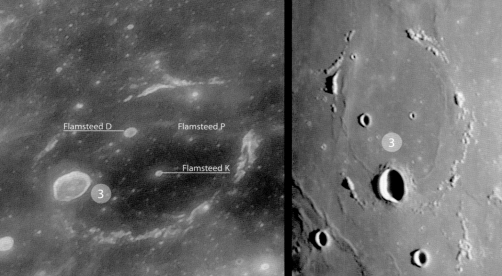

Letronne 10.8°S, 42.5°W ❶

Letronne is a semicircular relief feature, open to the north, and lying north of Gassendi. It is the remnant of a lava-flooded crater, 116 km across, and appears like a bay of Oceanus Procellarum. Morphologically, Letronne resembles the terrestrial, Chicxulub meteorite crater on the Yucatan peninsula in Mexico. A similar structure is the crater Le Monnier in Mare Serenitatis.

On the floor of Letronne three small mountain peaks (the remnants of the central mountain) are visible. Northwest of these lies a craterlet surrounded by a bright halo. Near the inner eastern wall lies Letronne B (5 km), which is also surrounded by a bright halo. The mare wrinkle-ridge system Dorsa Rubey (10.0°S, 42.0°W) runs north-south and ends inside Letronne at the remnants of the central mountain peak. The overall length of Dorsa Rubey amounts to about 100 km.

Wichmann 7.5°S, 38.1°W ❷

Wichmann is a small, circular, and very bright crater, 10 km across. It lies at the southeastern end of a large semicircular formation, named Wichmann R, which is undoubtedly the remnants of the wall of a crater that has been completely flooded by lava from the south. South of Wichmann lies a small mountain and directly east of that the craterlet Scheele (formerly named Letronne D), which has a diameter of only 4 km.

Flamsteed 4.5°S, 44.3°W ❸

A crater, 20 km in diameter. It lies near the southern section of the wall of the ghost crater Flamsteed P. Flamsteed P is a partially breached elliptical ring of mountain ridges, with a diameter of about 110 km. It is interpreted as being an old impact crater, which was submerged by lava when the Procellarum Basin was flooded. Under grazing illumination, a very low feature, like a mare wrinkle ridge, is visible within Flamsteed P, and which more-or-less follows the course of the crater wall. Within Flamsteed P lie the small craters Flamsteed D (6 km) and K (4 km).

The lunar probe Surveyor 1 made a soft landing on 2 June 1966 near the northeastern inner crater wall (the highest spot), and returned more than 11 000 images from there. The probe was NASA's first successful soft landing on the Moon. It was an important mission in preparation for the Apollo programme.

Winthrop 10.7°S, 44.4°W ⑤

The remnant of a crater, 17 km in diameter, completely submerged by lava, which lies on the western wall of Letronne.

Billy 13.8°S, 50.1°W ⑥

Billy is a crater 45 km in diameter. Even in larger telescopes the crater floor shows no form of structure. The lava that has flooded the crater is conspicuously dark and resembles the lava in the crater Crüger. The rille Rima Billy runs south for about 70 km, parallel to a mountainous ridge, which splits at the end rather like a spanner or wrench. Rima Billy is a linear rille and, with a maximum width of 2 km is observable even in medium-sized telescopes. Lying west of Rima Billy is the rille system Rimae Zupus, and east of it part of Rimae Mersenius.

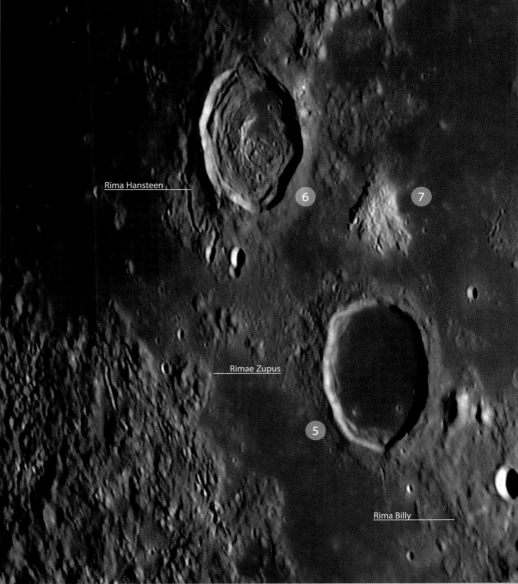

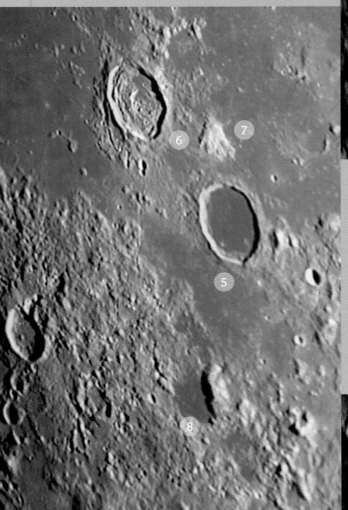

Hansteen 11.5°S, 52.0°W ⑦

Hansteed, lying northwest of Billy, presents a completely different appearance from Billy. The crater has a diameter of 44 km, and the floor is rough, furrowed and exhibits fracture zones. Directly west of the crater rim lies a short, narrow rille, Rima Hansteen (length 25 km, width 3 km).

Mons Hansteen 12.1°S, 50.0°W ⑧

A triangular, isolated mountain, with a base measuring approximately 30 × 30 km. It lies somewhat to the east, but half-way between Hansteen and Billy. When the angle of illumination is increasing, it appears very bright. Its origin remains unknown.

Zupus 17.2°S, 53.0°W ⑧

The remnant of a lava-flooded, elliptical crater, 38 × 47 km across. The crater's floor is as dark as that of Billy. The eastern crater wall appear linear and is higher than the western wall. Between Zupus C and Hansteen lie Rimae Zupus, a narrow rille system (120 km, 1.5 km wide, 15.0°S, 53.0°W). Observation of these rilles requires a large instrument.

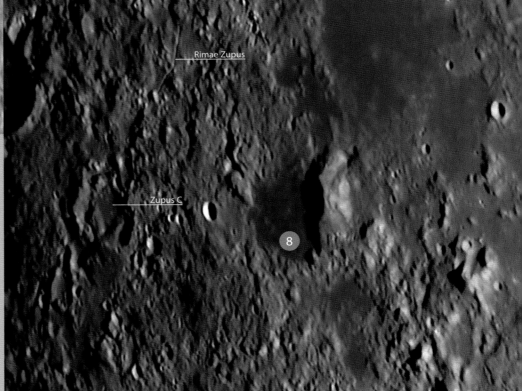

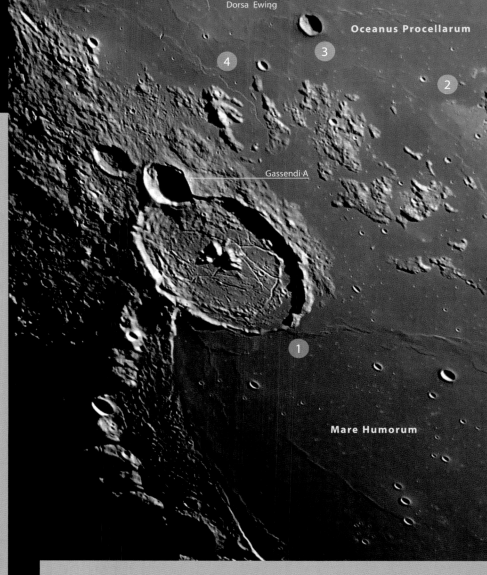

Gassendi 17.6°S, 40.1°W ①

The crater Gassendi is a prominent example of a crater that was altered after the impact, over a long period of time by post-impact volcanism and lava flows. It lies on the northern edge of Mare Humorum and has a diameter of 101 km and a depth of about 1.9 km. Dating the formation of the basin from measurements made by the European Moon probe Smart-1 gives a value of about 3.6 billion years. When the lava flows flooded the Humorum impact basin, large parts of the crater floor were also filled with lava. The crater wall and a few peaks of the central mountains are still visible. The whole floor of the crater is crossed by Rimae Gassendi, a complicated system of rilles, fissures and grabens. Located on the floor are the craterlets Gassendi P (2 km, in the northwest) and Gassendi M (3 km, in the southeast). The northern crater wall is breached by Gassendi A (33 km). The western wall shows a triangular landslip (similar to the one in the crater Plato).

The Helmet 16.7°S, 31.5°W ②

The Helmet is a low rise in the lunar crust with a diameter of about 60 km and a very low height of less than 200 metres. The structure is classified as a volcanic megadome. The surface is covered in rocky ridges and exhibits a few small craters. On colour-coded multispectral images of the region the structure stands out strinkingly like a 'beacon' from the surrounding mare lava through its brilliant colours. The area absorbs ultraviolet and blue light and is strongly reflective in the red and near-infrared spectral regions. Whereas the brownish areas contain rocks with a high iron content, the bluish ones are enriched in titanium dioxide.

'The Helmet' is not an official name for this area. The origin of the name is uncertain. In some sources it is ascribed to the crew of Apollo 16 during one of their passes, whereas in others the name of Pieters, an American planetologist is cited. Successful observation, because of its low height, requires grazing illumination at sunrise or sunset over the area.

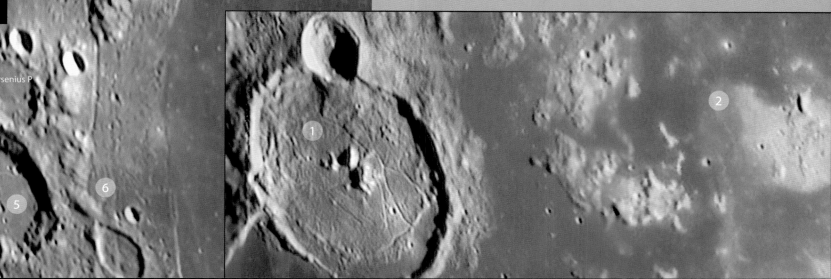

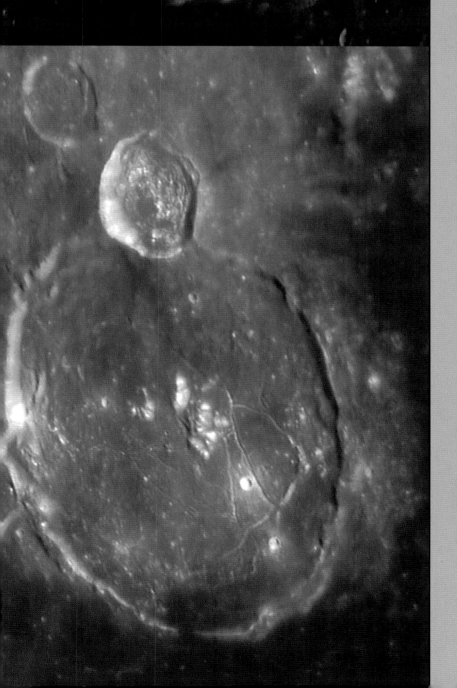

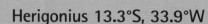
Gassendi A

Gassendi P

Gassendi M

1

Herigonius 13.3°S, 33.9°W   3

Herigonius is a rather insignificant crater, 15 km in diameter. It lies directly on top of one section of the mare wrinkle ridge system Dorsa Ewing (11.0°S, 38.0°W). A volcanic dome lies northeast of the crater.

Rimae Herigonius 13.0°S, 37.0°W   4

This rille system consists mainly of a long sinuous rille with many branches, which forks into a 'Y' at the northern end. The rille runs in a meandering fashion west from Gassendi and Letronne between Mare Humorum and Oceanus Procellarum, for an overall length of 100 km. A notable feature is the rectangular depression between the Rimae Herigonius and Cassini A. The floor of the depression appears rough. The rille is only about 1 km wide, and its observation requires a larger telescope and appropriate lighting conditions.

Mersenius 21.5°S, 49.2°W   5

Mersenius is a striking, large, lava-flooded crater, 84 km in diameter, lying west of Mare Humorum. The crater floor appears slightly convex. Mersenius possibly exhibits an intermediate stage between a normal complex crater and an FFC crater. A small, nameless crater chain lies on the crater's floor, oriented roughly north-south. Bordering on the crater wall on the northeast is Mersennius P with a diameter of 42 km.

Rimae Mersenius 20.0°S, 45.0°W   6

Rimae Mersenius is a system of wide linear rilles with a length of almost 300 km, which is visible even in small telescopes. It lies east of the crater Mersenius. The rilles probably arose through tension and shear stresses generated by the subsidence of the large lava deposit in the Humorum Basin.

Mare Nubium

The Helmet

Bullaldus

Gassendi

7

9

Campanus

2

1

8

Palus Epidemiarum

3

Mare Humorum 24.0°S, 39.0°W ①

Like all the other lunar maria Mare Humorum ('Sea of Humours')
is also an impact basin flooded in the centre with mantle lava.
With a diameter of about 380 km (that is, of the lava-covered
area: the actual basin is larger), it is one of the smaller maria,
and is roughly comparable with Iceland in area. The Humorum
Basin is also associated with a gravitational anomaly (a mascon).

The formation of the basin has been dated to the Nectarian period, approximately 3.9 to 3.8 billions years ago. Flooding with
lava occurred much later. The lava layer is over 3 km thick in the
central region. After the lava cooled and with the cessation of
the pressure from the lava rising from the mantle, the lava layers collapsed under their own enormous weight and deformed
the lunar crust beneath, thus creating the mare wrinkle ridges,
fracture zones and escarpments around the edge of the basin.
When the lava surface sank, previously existing craters and crater
floors around the edge of Mare Humorum were therefore tilted
in the direction of the centre. An area of Dark Mantle Deposit
(DMD) lies on the floor of the mare, with deposits of dark, glas-
sy, volcanic ashes, and which is crossed by a rille, whose origin
is uncertain.

Promontorium Kelvin 28.8°S, 33.6°W ②

Cape Kelvin is a free-standing large mountainous ridge. South
of it lies Rupes Kelvin (28°S, 33°W), a 150-km-long scarp. Rupes Kelvin is a portion of the inner wall of the Humorum Basin.

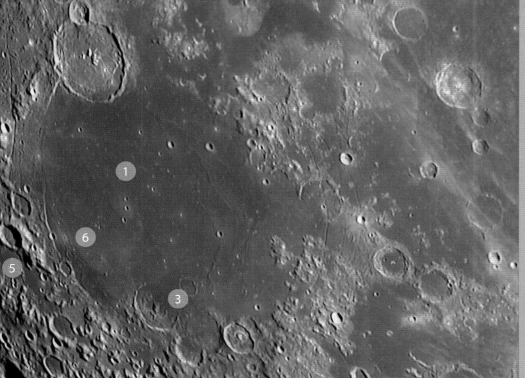

Doppelmayer 28.5°S, 41.4°W ③

A heavily eroded crater, 64 km in diameter, almost open to Mare Humorum in the northeast. The tilt of the crater towards the centre of the Humorum Basin is clearly visible. Doppelmayer's central mountain shows an almost pyramid-like shape under certain lighting conditions. Northeast of the crater lie Rimae Doppelmayer, following the curvature of the edge of the mare. They cross an area with ash deposits of very dark mantle material (a DMD area), and have an overall length of about 160 km and are mirrored by the Rimae Hippalus on the eastern side of the Humorum Basin.

Vitello 30.4°S, 37.5°W ④

Vitello is a crater 42 km across on the southern edge of Mare Humorum. The crater is shallow, with the walls reaching just 1.7 km above the floor. Its central peak is surrounded by a C-shaped gap. West of Vitello lies the crater Lee (77 km, 30.7°S, 40.7°W).

Liebig 24.3°S, 48.2°W ⑤
Rupes Liebig 25.0°S, 46.0°W ⑥

A 37-km-diameter crater with a few craterlets on the floor. Rupes Liebig runs as an impressive escarpment along the western edge of Mare Humorum with an overall length of 180 km, interrupted at about half-way by the craters Liebig G (20 km) and Liebig F (9 km).

Loewy 22.7°S, 32.8°W ⑦
Puiseux 27.8°S, 39.0°W ⑧

Two lava-flooded craters, 24 km in diameter. In between lies a complex system of unnamed mare wrinkle ridges, which become strikingly visible under low solar illumination. Puiseux is almost completely submerged, with only the highest parts of the crater wall emerging from the lava of Mare Humorum.

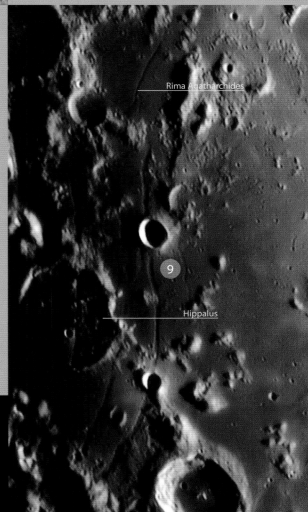

Rimae Hippalus 25.0°S, 29.0°W ⑨

A spectacular system, consisting of three rilles, all about 2 km wide, that stretch in an arc over a length of just under 200 km. The curvature of the rilles follows approximately the outer periphery of Mare Humorum. The westernmost of the three rilles runs right through the 58-kilometre wide, completely destroyed, crater remnant of Hippalus. Rimae Hippalus are the result of subsidence rifting of the enormous mass of lava in Mare Humorum after the lava cooled. The rilles appear most striking when illuminated near the terminator.

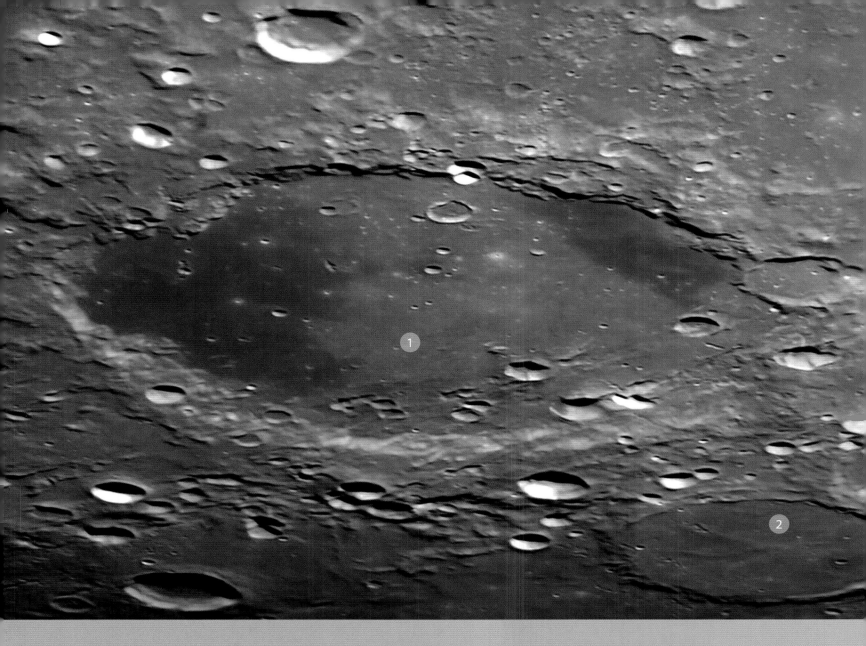

Schickard 44.3°S, 55.3°W

Schickard is a very large, old crater, 206 km in diameter, with a relatively smooth and level floor. Under certain lighting conditions the crater's floor appears convex, bulging upwards slightly. Following its formation, the crater was re-structured by the Mare Orientale impact, almost 1200 km away. Bright ejecta from the Orientale impact may be observed on the crater's floor and the western wall exhibits radial, linear furrows and graben, oriented to the centre of Mare Orientale. Long after the Orientale impact, the northern and southern portion of the crater's floor were flooded with dark lava.

Wargentin 49.6°S, 60.2°W

Wargentin is one of the rare plateau craters, and is filled to the rim with dark lava. The crater's floor lies about 400 m above the surrounding terrain and is crossed by a mare wrinkle ridge. The network of the range of hills possibly marks the location of the former central mountains and ridges. A craterlet, surrounded by a dark halo, lies on the southern portion of the crater's floor. Because of its location near the limb, Wargentin is best observed at favourable libration angles.

Phocylides 52.7°S, 57.0°W

A large, lava-filled old crater, 120 km in diameter, and with 2.1 km high crater walls. The eastern wall is completely eroded. The crater Phocylides C, 46 km across, directly adjoins it on the northwest.

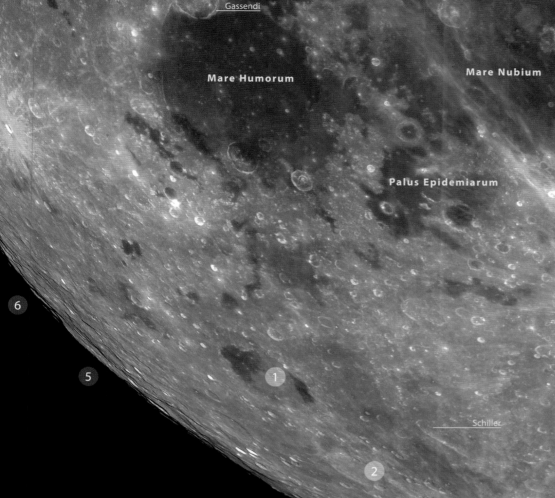

Billy

Gassendi

Mare Humorum

Mare Nubium

Palus Epidemiarum

Schiller

⑥

⑤

① ② ⑦

Inghirami ④
47.5°S, 68.8°W
A prominent crater, 91 km in diameter. It was also altered by the Orientale impact. Its floor shows radial ridges and valleys directed towards the centre of the Orientale Basin.

Piazzi ⑤
36.5°S, 67.9°W
Lagrange ⑥
32.3°S, 72.8°W
Two large, old, complex craters, largely destroyed by the Orientale impact, with diameters of 100 km and 160 km, respectively. The floor of Lagrange is divided in two by an enormous ridge.

Nasmyth ⑦
50.5°S, 56.2°W
A lava-filled crater, 76 km in diameter. Portions of the western wall are overlapped by Phocylides and Wargentin.

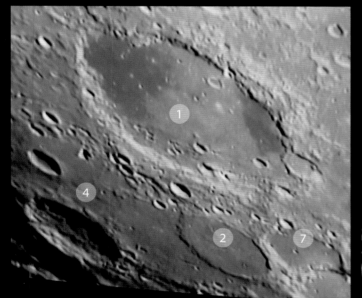

① ④ ② ⑦

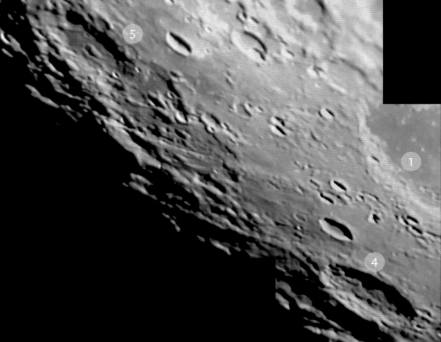

⑤ ① ④

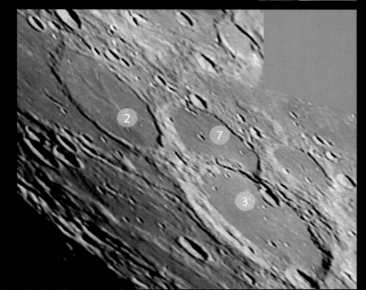

② ⑦ ③

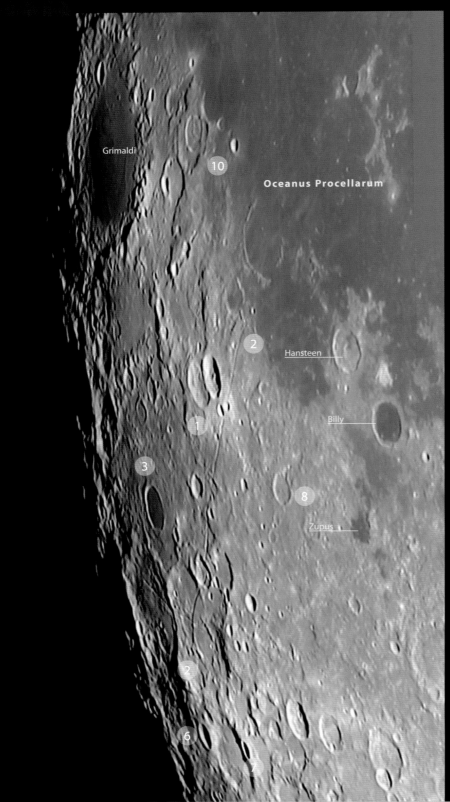

Grimaldi

10

Oceanus Procellarum

2

Hansteen

1

Billy

3

8

Zupus

2

6

7

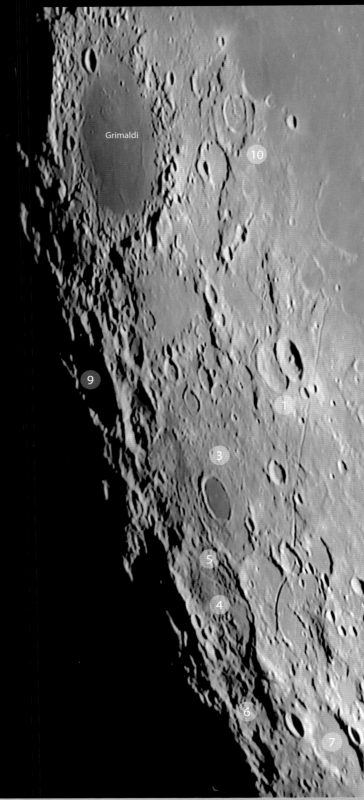

Grimaldi

10

9

1

3

5

4

6

7

Sirsalis 12.5°S, 60.4°W
Rimae Sirsalis 15.7°S, 61.7°W

1
2

Sirsalis is a crater 42 km across, with a central peak. It overlaps the crater Sirsalis A, 49 km across, which was formerly known as Bertaud. Sirsalis J, directly adjoining Sirsalis to the southeast, is a small crater, 12 km in diameter and the centre of a small ray system. The 3-km wide rille Rima Sirsalis run north-south on the eastern side of Sirsalis, and which is easily seen even in small telescopes. Rima Sirsalis is a fracture zone, possibly caused by the Orientale impact. To the south, Rima Sirsalis merges directly into the rille system Rimae Darwin. The widest of the Sirsalis rilles has an overall length of more than 300 km.

A narrow ridge begins at the northern wall of Sirsalis and forms the eastern wall of the crater Sirsalis Z (91 km), which casts a bizzare triangular shadow around the 12th day after New Moon. The floor of Sirsalis Z is smooth and level, and a distant portion of Rimae Grimaldi ends inside the northeastern crater wall.

Sirsalis E (8.0°S, 56.0°W) is a crater, 80 km across, that is almost completely flooded by the Oceanus-Procellarum lavas. Only the western wall and a small section of the eastern crater wall are visible.

Crüger 16.7°S, 66.8°W

Crüger is a crater, 45 km in diameter, flooded with a very dark lava. The crater's floor appears completely flat and level. Large telescopes reveal a craterlet. Directly to the north lies Lacus Aestatis ('Lake of Summer') an irregularly shaped lava plain about 90 km across. This lava, too, is very dark in its tint.

Darwin 20.2°S, 69.5°W **4**
Rimae Darwin 20.0°S, 67.0°W **5**

Darwin is a large, very complex and almost completely destroyed crater, about 120 km in diameter. A group of smaller craters covers the southern portion of the crater's floor. The northern part of the floor is crossed by Rimae Darwin. The rille system has an overall length of 280 km, with a mean width of between 1 and 2 km. Between the rilles lies a large, peculiarly shaped hill (perhaps a lunar dome), but which is visible only under grazing illumination.

Lamarck 22.9°S, 69.8°W

Adjoining Darwin to the south is the similarly extremely broken-down crater Lamarck, with a diameter of 110 km. The southern wall of Darwin and the northern wall of Lamarck overlap.

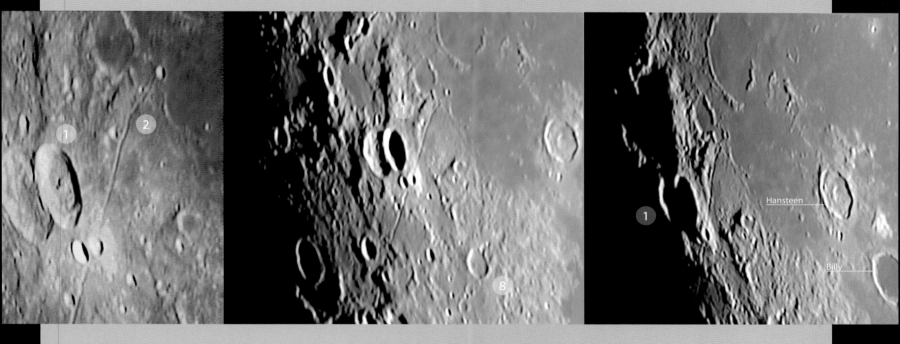

Byrgius 24.7°S, 65.3°W **7**

Byrgius is a crater 87 km in diameter. The floor is smooth and level, and is covered with bright ejecta from the crater Byrgius A (19 km), which lies on the eastern crater wall. Under high solar illumination, Byrgius A exhibits a ray system.

Fontana 16.1°S, 56.6°W

A crater, 31 km in diameter, with a level floor. Under grazing illumination larger telescopes show a low hill on the crater's floor. The northeastern wall of the crater is heavily eroded.

Rocca 12.7°S, 72.8°W **9**

A largely destroyed crater, 90 km in diameter, with a rough, furrowed floor.

Damoiseau 4.8°S, 61.1°W **10**

Damoiseau is a relatively shallow crater, 36 km in diameter, which lies in a larger crater, Damoiseau M (54 km). The floor of Damoiseau is uneven, rough, and exhibits a few ridges. A valley-like structure east of Damoiseau has been named as Damoiseau K. One of the main rilles of Rimae Grimaldi ends at the southwestern wall.

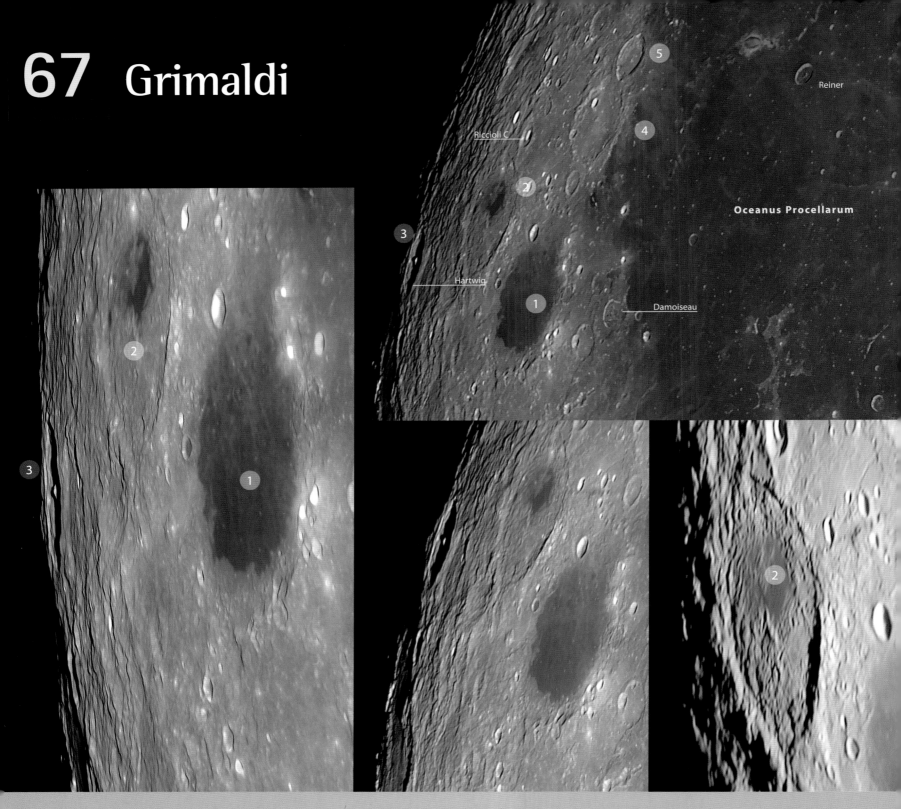

Grimaldi 5.5°S, 68.3°W ①

Grimaldi, because of its size, must be counted as an impact basin (similar to Mare Humboltianum), whose lava-flooded central portion is bounded by an inner wall that is about 200 km in diameter. The outer wall, which is only partially remaining, has a diameter of over 400 km. Grimaldi may thus be classed as a multi-ring basin.

Running parallel to the southeastern portion of the inner crater wall are Rimae Grimaldi, with an overall length of 230 km (with a maximum width of 2 km). They are narrow linear rilles, probably produced by the tension forces when the lava flows contracted as they cooled. In the northern portion of the crater's floor there is a large volcanic dome with a summit crater. A few additional domes may be observed with larger telescopes under favourable lighting and seeing conditions. Grimaldi is associated with a mascon, a gravitational anomaly. Because Grimaldi is near the western limb of the Moon, observation of the Grimaldi Basin and its rille system requires favourable libration angles.

Riccioli 3.3°S, 74.6°W ②

Riccioli is a large complex crater, 145 km in diameter. The Rimae Riccioli lie in the southern portion of the crater's floor, and stretch for a total length of over 400 km. They are linear rilles with a maximum width of 1.5 km. The northern portion of the crater floor is lava-flooded. Riccioli C, lying directly north of Riccioli, is a crater 31 km across with a further impact crater on its floor. It is a rare example of a centrally located double impact. Because of the extreme location near the western limb of the Moon, observations succeed best under favourable libration conditions. Images from unmanned lunar probes clearly show that Riccioli has been subsequently modified by ejected material from the Orientale impact.

The crater commemorates Giovanni Battista Riccioli. Riccioli (1598–1671) was an Italian theologian and Moon observer. He was the first to name craters systemmatically after astronomers, scientists and philosophers, and is the originator of lunar nomenclature.

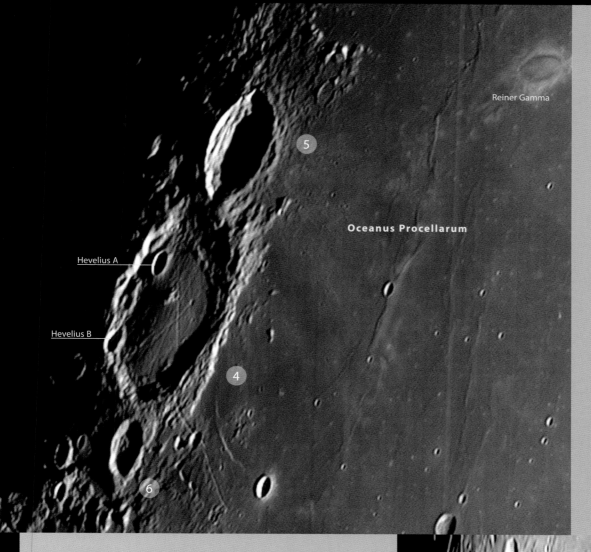

Reiner Gamma

Oceanus Procellarum

Hevelius A

Hevelius B

Schlüter 5.9°S, 83.3°W 3

Schlüter is a crater, 89 km across, with central mountains and terraced inner crater walls. Directly neighbouring it to the east is the crater Hartwig (6.1°S, 80.5°W), with a diameter of 79 km, only slightly smaller than Schlüter.

Hevelius 2.2°N, 67.6°W 4

Hevelius is a medium-sized, but nevertheless conspicuous large crater with a diameter of 115 km and some remnants of central mountains. The crater's floor is broken in a cross-shape by Rimae Hevelius (with a total length of about 180 km, and a maximum width of about 1.5 km). In addition, on the crater floor and on the western wall, there are a few small and large craters visible (Hevelius A, B, G, E, F and H). The largest are Hevelius A (on the crater floor) and Hevelius B (on the western wall) with diameters of about 14 km.

Cavalerius 5.1°N, 66.8°W 5

North of Hevelius, and directly adjacent, is the crater Cavalerius with a diameter of 57 km and 3 km deep. The crater has terraced inner walls and exhibits the peak of a central mountain on the crater floor.

Lohrmann 0.5°S, 67.2°W 6

A normal crater, 30 km in diameter, with a smooth floor and a few low hills within.

Hedin 2.0°N, 76.5°W 7

A medium-sized complex crater, 150 km in diameter – also extensively destroyed by the Orientale impact. The floor of Hedin is covered in ejecta from the Orientale impact.

Olbers 7.4°N, 75.9°W 8

A crater, 74 km in diameter, and with an extremely smooth floor. The northwestern wall is breached by the crater Glushko, which was designated Olbers A on earlier Moon maps. Glushko is a very young crater and the centre of bright rays, with cover Oceanus Procellarum and end at the crater Schiaparelli. The ray system appears very bright against the dark mare lava.

Glushko

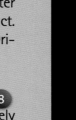

Mare Orientale 19.0°S, 93.0°W ①

Mare Orientale ('Eastern Sea') is the lava-flooded, nearly circular portion of a very large impact basin. Whereas the diameter of Mare Orientale only amounts to 320 km, the basin measures about 930 km in diameter. The mare is surrounded by two ramparts, which enclose the lava surface with concentric rings, and which were created by seismic shock waves from the impact. These circular mountain ranges are known as Montes Rook (the Rook Mountains, named after Lawrence Rook, an English astronomer and observer of Jupiter's moons (1622–1666) and Montes Cordillera, where Montes Cordillera form the outer rampart.

Only the eastern outer regions of Mare Orientale are visible – at favourable libration angles – including:
the eastern edge of Montes Cordillera (20.0°S, 80.0°W),
the eastern edge of Montes Rook (20.0°S, 83.0°W),
Lacus Autumni ('Lake of Autumn', 14.0°S, 82.0°W), and Lacus Veris ('Lake of Spring', 13.0°S, 87.0°W).
Lacus Autumni is a narrow lava area, oriented north-south, about 250 km long at the inner edge of Montes Cordillera. Lacus Veris consists of two neighbouring, similarly narrow, lava areas with a total length of about 550 km at the inner edge of Montes Rook.

The ejecta from the Orientale impact may be traced out to distances of 1200 km, for example on the floor of the crater Schickard.

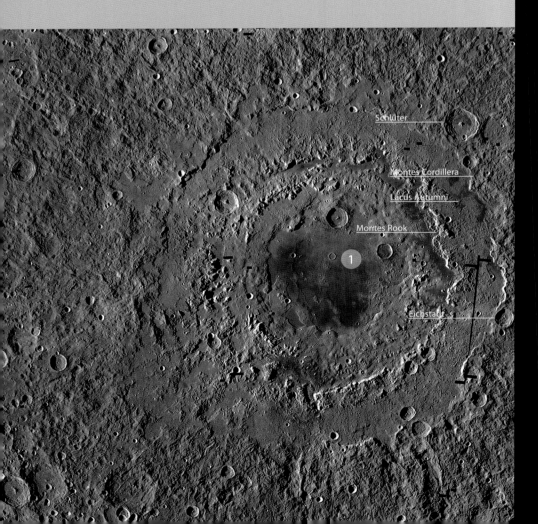

Because of libration effects, only about 10 per cent of the Moon's farside may be observed from Earth. About 40 per cent of the Moon's surface, however, remains permanently hidden.

The first, rather blurry views of the Moon's farside were obtained by the Soviet Moon probe Luna 3 in October 1959. Although they may have been unsharp and blurred, these pictures showed, however, that the surface of the farside differed greatly from the nearside. On the farside, for example, there are only four small maria, and a meagre half-dozen large craters whose floors are lava-flooded.

The ratio of the areas of mare surface to the crater-saturated highlands on the Moon's nearside amounts to about 31.2 per cent to 68.8 per cent, and thus roughly 1:2.2. On the farside the ratio is, however, 2.6 per cent to 97.4 per cent, so roughly 1:37.5. Almost the whole of the farside of the Moon lies well above the average level of the surface.

One reason for this is undoubtedly the fact that the crust on the farside, at up to 140 km thick, is nearly double the thickness of the crust on the nearside. What led to this striking difference is still under discussion by planetologists and geologists.

But the fact is that impact events would have had to have been significantly more violent on the farside to break through the crust to produce cracks and fissures through which lava could flow. As such, the distance that the mantle lava would have to rise vertically would be twice that on the nearside.

Montes Leibniz
Under favourable solar illumination (when the Moon's age is about 9.5 days) and suitable libration, a series of high mountain peaks may be observed at the southern limb of the Moon, which stand out against the black background of the sky.

The first observations of these mountains were made by Johan Hieronymus Schröter in December 1789. He named the mountain peaks after the Hannoverian universal scholar Gottfried Wilhelm Leibniz and designated the individual peaks Leibnitz a to Leibnitz e. Schröter also produced the first measurements of the height of Leibniz b, which gave a value of 5.5 km. Later observers replaced Schröter's designations a to e with Greek, lower-case letters, beginning with Leibniz Alpha. In 1935, the IAU officially designated the range as Montes Leibniz and the individual mountain peaks with Greek letters, but in 1971 the names were finally deleted, without replacement.

South Pole-Aitken Basin
Even before the first pictures of the Moon's farside, individual researchers suggested that the mountain peaks might be the outer ramparts of a very large impact basin on the lunar farside. Later, higher-resolution images by the Lunar Orbiter probes, however, showed no such lava-filled impact basin, but only a collection of larger craters with lava-covered floors.

Precise laser height measurements by the Apollo missions and subsequent lunar probes (such as Clementine) confirmed the hypothesis of Hartmann and Kuiper; an enormous basin actually existed on the Moon's farside, with a depth of as much as 8 km (although not, in general, flooded by lava) measured relative to the average level of the surrounding terrain.

The Leibniz mountains are a portion of the outer wall of the basin. On the nearside it runs south of the craters Boussingault, Schomberger, Newton and le Gentil. According to modern measurements, Leibniz Beta reaches a height of about 7 km, and the difference in height between the lowest and the highest points of the South Pole-Aitken Basin amounts to approximately 15 to 16 km. With a diameter of about 2400 km it is the second largest basin structure in the inner Solar System, and is exceeded only by the Hellas Basin on Mars.

South Pole-Aitken is certainly a very old basin (probably created 3.9 billion years ago), because the whole structure is covered in innumerable large and small craters, which were formed later. It stretches from the Moon's South Pole to the Aitken crater, and exhibits only a few places that were flooded with lava, such as the great crater-like features Leibnitz, von Kármán, Apollo, Poincaré – a lava-filled area north of the crater Bose – and Mare Ingenii ('Sea of Ingenuity', diameter of about 300 km). Apollo is an impact basin with a diameter of about 500 km.

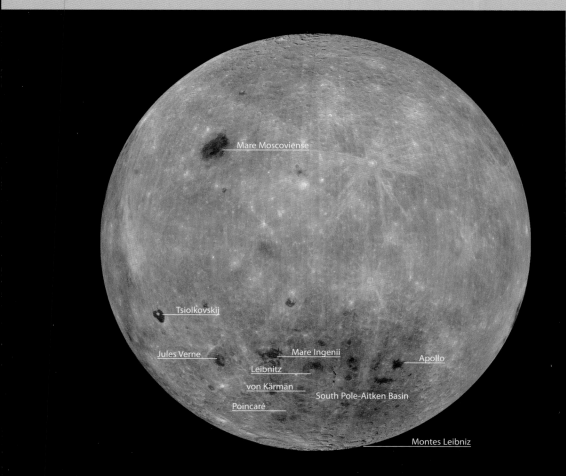

Mare Moscoviense
Tsiolkovskij
Jules Verne
Mare Ingenii
Apollo
Leibnitz
von Kármán
South Pole-Aitken Basin
Poincaré
Montes Leibniz

Glossary

albedo   The reflectivity of a body. Coal has a low albedo, and a sheet of white paper a high albedo. At Full Moon, the average albedo of the surface amounts to only 7 per cent. The brightest areas, such as the ray system of Aristarchus are about 20 per cent.

anorthosite   A light-coloured surface rock in the lunar highlands.

apogee   The point on the Moon's orbit that is most distant from Earth.

Blue Moon   1) A rare occurrence, when the Moon (or the Sun) appears blue, normally caused by the presence of aerosols (particularly particles derived from forest fires) in the atmosphere. Hence the saying 'Once in a Blue Moon'.
2) In recent years, and particularly in North America, this term has been applied to the second Full Moon in a particular calendar month. Such an occurrence is relatively rare: on average in 100 years there are 41 occasions when the Full-Moon phase occurs twice in equal calendar months, specifically in months with 31 days.

breccia   A geological term for a rock consisting of fragments, sometimes of different compositions, which have been welded together by a high-temperature process. In the case of the Moon, breccias have frequently been formed by the high temperatures and pressures present during an impact.

Cassini's White Cloud   The name for a bright area in Deslandres (also known as 'Cassini's bright spot'), where G.D. Cassini claimed to have seen a 'white cloud'.

colongitude   The lunar longitude at which the Sun is rising. The position of the terminator, measured westwards from the zero meridian from 0° to 360°.

Dark Mantle Deposit   An area covered in a layer of dark ash created by eruptive volcanism.

Earthshine   Ashen-coloured light reflected and scattered by the Earth's atmosphere, which illuminates regions of the Moon that are not directly sunlit.

ejecta blanket   An area covered with a continuous layer of material ejected during the formation of an impact crater.

ecliptic   The projection of the Sun's orbit onto the celestial sphere.

erosion   The 'weathering' and ageing processes that affect lunar rocks, primarily caused by the bombardment by micrometeorites, electrons and protons from the solar wind, and cosmic rays.

extrusion   The emission of lava onto the surface. (In the case of the Moon, most lava flows are of very fluid magma.)

FFC crater   A Floor Fractured Crater, with an extensive and prominent rille system on the crater floor.

ghost crater   A very old crater that has been almost completely covered by subsequent lava flows or impact ejecta.

Genesis Rock   The oldest piece of lunar rock, brought back to Earth by the Apollo 15 mission. Its formation has been dated to 4.5 billion years ago.

IAU   International Astronomical Union, which, among other functions, is the sole organisation responsible for naming features on the Moon, planets, and other celestial bodies.

Imbrium Sculpture   The term for linear relief features that extend radially from the Imbrium Basin, and formed by the impact that created the Imbrium Basin.

intrusion   The injection of lava into layers of rock. The lava may fill cracks and small voids, but often penetrates between existing layers of overlying rock, lava flows, ash deposits, etc. If the intrusion raises higher layers, the result may be a volcanic uplift dome.

KREEP   The term for lunar rock that is rich in potassium (K), rare earth elements (REE), and phosphorus (P).

LHB   The abbreviation for Late Heavy Bombardment. It is the period after the formation of the Moon when the planets of the inner Solar System and the lunar surface were subjected to intense bombardment by planetesimals remaining after the formation of the planets, leading to the creation of lunar craters and basins.

libration   The rocking of the Moon in its orbit around the Earth, which allows more than 50 per cent of the surface to be observed from Earth.

LIDAR   Light Detection And Ranging (normally using a laser source). A LIDAR instrument was employed on the Clementine lunar probe to produce a high-resolution map of the altitude of the lunar surface.

LLR   Lunar Laser Ranging, a method using laser light that allows the distance of the Moon from Earth to be determined with a high degree of accuracy.

LTP   Abbreviation for Lunar Transient Phenomena, a term used to describe observations of short-term (fleeting) changes on the lunar surface. Also known as TLP (Transient Lunar Phenomena).

magcon   An acronym for Magnetic Concentration, a magnetic-field anomaly.

mascon   An acronyn for Mass Concentration, an anomaly in the Moon's gravitational field.

megaregolith   The extremely thick layer of regolith in the Moon's highland regions.

morphology   The appearance or actual (geological) structure of a surface region.

multi-spectral images   Images of the lunar surface, taken in various wavelengths. From these images the mineralogical composition of lunar rocks and lavas may be determined.

perigee   The point on the Moon's orbit closest to the Earth.

planetesimal   The term for large or small bodies, similar to those of the minor planets (asteroids), remaining from the time of the formation of the Solar System, and which did not form into planets.

radiometric dating   A method of determining the age of rocks by measurement of the half-lives of naturally occurring radioactive elements.

regolith   A layer of loose rocky material, that has formed from the underlying original rock through weathering and erosion.

Schröter's Law   A rule of thumb, advanced by Johann Hieronymus Schröter, that the volume of a crater's walls corresponds approximately to the volume of the crater's interior.

sidereal month   The average period for the Moon to make one complete orbit of the Earth, relative to the stars (27.3217 days), and thus shorter than the synodic month.

stratigraphy   The most important method of relative dating of the succession of layers in sedimentary rocks.

superposition   The method of stratigraphy, as applied to the Moon. The relative dating of surface features is determined on the principle of overlapping. (For example: a crater filled with lava must have been formed before the lava flow occurred.)

synodic month   The average period between two identical phases of the Moon such as between successive Full Moons (29.53 days).

terminator   The dividing line between the illuminated and non-illuminated portions of the Moon's surface, and thus between day and night. The morning terminator traces sunrise, and the evening terminator, sunset. Because of the absence of an atmosphere, there is practically no scattered light, and consequently no twilight effects. So the dividing line is sharply defined. Near the terminator, the Sun's illumination is at a very shallow angle and even the smallest surface irregularities cast long shadows and thus become apparent in three dimensions.

USGS   Abbreviation for United States Geological Survey, which maintains a database of lunar nomenclature.

Index of lunar features

Name	Longitude/Latitude	Chapter
Abenezra	21.0°S, 11.9°E	15
Abulfeda	13.8°S, 13.9°E	15
Agrippa	4.1°N, 10.5°E	33
Albategnius	11.7°S, 4.3°E	36
Alexander	40.3°N, 13.5°E	24
Aliacensis	30.6°S, 5.2°E	39
Alpetragius	16.0°S, 4.5°W	37b
Alphonsus	13.4°S, 2.8°W	37b
Anaxagoras	73.4°N, 10.1°W	57
Anaximenes	72.5°N, 44.5°W	56
Arago	6.2°N, 21.4°E	19b
Archimedes	29.7°N, 4.0°W	30
Aristarchus	23.7°N, 47.4°W	58
Aristillus	33.9°N, 1.2°E	25
Aristoteles	50.2°N, 17.4°E	23
Aryabhata	6.2°N, 35.1°E	19a
Arzachel	18.2°S, 1.9°W	37b
Atlas	46.7°N, 44.4°E	5
Atlas A	45.3°N, 49.6°E	5
Autolycus	30.7°N, 1.5°E	25
Azophi	22.1°S, 12.7°E	15
Babbage	59.7°N, 57.1°W	56
Bailly	66.5°S, 69.1°W	44
Beaumont	18.0°S, 28.8°E	13b
Berosus	33.5°N, 69.9°E	3
Bessel	21.8°N, 17.9°E	20a
Bianchini	48.7°N, 34.3°W	54
Biela	54.9°S, 51.3°E	11
Billy	13.8°S, 50.1°W	62
Birmingham	65.1°N, 10.5°W	57
Birt	22.4°S, 8.5°W	38
Blancanus	63.8°S, 21.4°W	42
Bode	6.7°N, 2.4°W	35
Boguslawsky	72.9°S, 43.2°E	11
Bohnenberger	16.2°S, 40.0°E	13b
Bonpland	8.3°S, 17.4°W	48
Boussingault	70.2°S, 54.6°E	11
Briggs	26.5°N, 69.1°W	60
Bruce	1.1°N, 0.4°E	35
Bullialdus	20.7°S, 22.2°W	47
Burckhardt	31.1°N, 56.5°E	3
Bürg	45.0°N, 28.2°E	22
Byrd	85.3°N, 9.8°E	57
Byrgius	24.7°S, 65.3°W	66
C. Herschel	34.5°N, 31.2°W	53b
Calippus	38.9°N, 10.7°E	24
Campanus	28.0°S, 27.8°W	45
Capella	7.5°N, 35.0°E	17
Cardanus	13.2°N, 72.5°W	60
Casatus	72.8°S, 29.5°W	44
Cassini	40.2°N, 4.6°E	26
Cassini's White Cloud		39
Catena Abulfeda	17.0°S, 17.0°E	15
Catena Krafft	15.0°N, 72.0°W	60
Catharina	18.1°S, 23.4°E	16
Cauchy	9.6°N, 38.6°E	19a
Cavalerius	5.1°N, 66.8°W	67
Cayley	4.0°N, 15.1°E	33
Censorinus	0.4°S, 32.7°E	17
Chacornac	29.8°N, 31.7°E	21
Chevallier	44.9°N, 51.2°E	5
Ching-Te	20.0°N, 30.0°E	6
Clavius	58.8°S, 14.1°W	42
Cleomedes	27.7°N, 56.0°E	3
Conon	21.6°N, 2.0°E	31a
Copernicus	9.7°N, 20.1°W	51
Crüger	16.7°S, 66.8°W	66
Curtius	67.2°S, 4.4°E	41
Cyrillus	13.2°S, 24.0°E	16
da Vinci	9.1°N, 45.0°E	7
Daguerre	11.9°S, 33.6°E	13a
Dalton	17.1°N, 84.3°W	60
Damoiseau	4.8°S, 61.1°W	66
Daniell	35.3°N, 31.1°E	21
Darwin	20.2°S, 69.5°W	66
Davy	11.8°S, 8.1°W	37a
de la Rue	59.1°N, 52.3°E	4
Delambre	1.9°S, 17.5°E	18
Delisle	29.9°N, 34.6°W	53b
Descartes	11.5°S, 15.7°E	15
Deslandres	32.5°S, 5.2°W	39
Dionysius	2.8°N, 17.3°E	18
Diophantus	27.6°N, 34.3°W	53b
Doppelmayer	28.5°S, 41.4°W	64
Dorsa Serpentine	24.0°N, 25.0°E	20
Dorsum Heim	32.0°N, 29.8°W	53a
Dorsum Oppel	18.7°N, 52.6°E	2b
Dorsum Zirkel	28.1°N, 23.5°E	53a
Eddington	21.3°N, 72.2°W	60
Egede	48.7°N, 10.6°E	23
Einstein	16.3°N, 88.7°W	60
Encke	4.6°N, 36.6°W	59
Endymion	53.9°N, 57.0°E	4
Eratosthenes	14.5°N, 11.3°W	52
Eudoxus	44.3°N, 16.3°E	23
Euler	23.3°N, 29.2°W	53a
Fabricius	42.9°S, 42.0°E	12
Faraday	42.4°S, 8.7°E	40
Flammarion	3.4°S, 3.7°W	35
Flamsteed	4.5°S, 44.3°W	62
Fontana	16.1°S, 56.6°W	66
Fontenelle	63.4°N, 18.9°W	57
Fra Mauro	6.1°S, 17.0°W	48
Fracastorius	21.5°S, 33.2°E	13b
Furnerius	36.0°S, 60.6°E	9
Gambart	1.0°N, 15.2°W	48
Gardner	17.7°N, 33.8°E	6
Gassendi	17.6°S, 40.1°W	63
Gaudibert	10.9°S, 37.8°E	13b
Gauricus	33.8°S, 12.6°W	46
Gauss	35.7°N, 79.0°E	3
Geminus	34.5°N, 56.7°E	3
Gioja	83.3°N, 2.0°E	57
Goclenius	10.0°S, 45.0°E	8b
Godin	1.8°N, 10.2°E	33
Goldschmidt	73.2°N, 3.8°W	57
Gould	19.2°S,17.2°W	47
Grimaldi	5.5°S, 68.3°W	67
Gruemberger	66.9°S, 10.0°W	41
Gruithuisen	32.9°N, 39.7°W	55
Guericke	11.5°S, 14.1°W	48
Gutenberg	8.6°S, 41.2°E	8b
Gyldén	5.3°S, 0.3°W	35
Hahn	31.3°N, 73.6°E	3
Hainzel	41.3°S, 33.5°W	45
Hall	33.7°N, 37.0°E	21
Halley	8.0°S, 5.7°E	36
Hanno	56.3°S, 71.0°E	10
Hansteen	11.5°S, 52.0°W	62
Harpalus	52.6°N, 43.4°W	56
Hedin	2.0°N, 76.5°W	67
Helicon	40.4°N, 23.1°W	54
Hercules	46.7°N, 39.1°E	5
Herigonius	13.3°S, 33.9°W	63
Herodotus	23.2°N, 49.7°W	58
Herschel	5.7°S, 2.1°W	35
Hesiodus	29.4°S, 16.3°W	46
Hevelius	2.2°N, 67.6°W	67
Hipparchus	5.1°S, 5.2°E	36
Hommel	54.7°S, 33.8°E	11
Horrebow	58.7°N, 40.8°W	56
Horrocks	4.0°S, 5.9°E	36
Hortensius	6.5°N, 28.0°W	50
Huggins	41.1°S, 1.4°W	40
Inghirami	47.5°S, 68.8°W	65
Isidorus	8.0°S, 33.5°E	17
J. Herschel	62.0°N, 42.0°W	56
Jansen	13.5°N, 28.7°E	19b
Janssen	45.4°S, 40.3°E	12
Keldysh	51.2°N, 43.6°E	5
Kepler	8.1°N, 38.0°W	59
Kies	26.3°S, 22.5°W	47
Kircher	67.1°S, 45.3°W	44
Klaproth	69.8°S, 26.0°W	44
König	24.1°S, 24.5°W	47
Krafft	16.6°N, 72.6°W	60
Kunowsky	3.1°N, 32.5°W	59
La Condamine	53.4°N, 28.2°W	54
Lacus Felicitatis	19.0°N, 5.0°E	32
Lacus Mortis	45.0°N, 27.0°E	22
Lacus Spei	43.0°N, 65.0°E	3
Lacus Temporis	46.0°N, 57.0°E	4
Lacus Timoris	38.8°S, 27.3°W	45
Lade	1.3°S, 10.1°E	36
Lagrange	32.3°S, 72.8°W	65
Lalande	4.4°S, 8.6°W	35
Lamarck	22.9°S, 69.8°W	66
Lambert	25.8°N, 21.0°W	53a
Lamont	4.4°N, 23.7°E	19b
Langrenus	8.9°S, 61.1°E	9
Lansberg	0.3°S, 26.6°W	49
Lassell	15.5°S, 7.9°W	38
Le Monnier	26.6°N, 30.6°E	20b
Le Verrier	40.3°N, 20.6°W	54
Letronne	10.8°S, 42.5°W	62
Lichtenberg	31.8°N, 67.6°W	60
Lick	12.4°N, 54.7°E	2b
Liebig	24.3°S, 48.2°W	64
Lindenau	32.3°S, 24.9°E	14
Linné	27.7°N, 11.8°E	20a
Littrow	21.5°N, 31.4°E	6
Loewy	2.7°S, 32.8°W	64
Lohrmann	0.5°S, 67.2°W	67
Longomontanus	49.6°S, 21.9°W	43
Lubiniezky	17.8°S, 23.8°W	47
Lucian	14.3°N, 36.7°E	19a
Lyell	13.6°N, 40.6°E	19a
Lyot	50.2°S, 84.1°E	10
Macrobius	21.3°N, 46.0°E	7
Mädler	11.0°S, 29.8°E	16
Maestlin	4.9°N, 40.6°W	59
Maestlin R	3.5°N, 41.5°W	59
Maginus	50.0°S, 6.2°W	43
Mairan	41.6°N, 43.4°W	55
Manilius	14.5°N, 9.1°E	32
Marco Polo	15.4°N, 2.0°W	31b
Mare Australe	50.0°S, 93.0°E	10
Mare Cognitum	10.0°S, 23.0°W	49
Mare Crisium	17.0°N, 59.1°E	2
Mare Fecunditatis	5.5°S, 53.0°E	8
Mare Frigoris	56.0°N, 1.4°E	56
Mare Humboldtianum	56.8°N, 81.5°E	4
Mare Humorum	24.0°S, 39.0°W	64
Mare Imbrium	30.0°N, 20.0°W	53
Mare Insularum	7.0°N, 22.0°W	50
Mare Marginis	12.0°N, 88.0°E	1
Mare Nectaris	15.2°S, 35.5°E	13
Mare Nubium	20.0°S, 15.0°W	47
Mare Orientale	19.0°S, 93.0°W	68
Mare Serenitatis	28.0°N, 17.5°E	20
Mare Smythii	2.0°N, 87.0°E	1
Mare Spumans	1.0°N, 65.0°E	1
Mare Tranquillitatis	8.5°N, 31.5°E	19
Mare Undarum	7.0°N, 69.0°E	1
Mare Vaporum	13.0°N, 3.5°E	32
Marius	11.9°N, 50.8°W	61
Marth	31.1°S, 29.3°W	45
Mason	42.6°N, 30.5°E	22
Maupertius	49.6°N, 27.3°W	54
Maurolycus	42.0°S, 14.0°E	40
Mee	43.7°S, 35.3°W	45
Menelaus	16.3°N, 16.0°E	32
Mercator	29.3°S, 26.1°W	45
Mersenius	21.5°S, 49.2°W	63
Messala	39.2°N, 60.5°E	3
Messier	1.9°S, 47.6°E	8a
Messier A	2.0°S, 47.0°E	8a
Metius	40.3°S, 43.3°E	12
Meton	73.6°N, 18.8°E	57
Milichius	10.0°N, 30.2°W	50
Moltke	0.6°S, 24.2°E	18
Mons Blanc	45.0°N, 1.0°E	27
Mons Gruithuisen Delta	36.0°N, 39.5°W	55
Mons Gruithuisen Gamma	36.6°N, 40.5°W	55
Mons Hansteen	12.1°S, 50.0°W	62
Mons Herodotus	27.0°N, 53.0°W	58

Name	Coordinates	Page
Mons la Hire	27.8°N, 25.5°W	53b
Mons Moro	12.0°S, 19.7°W	48
Mons Penck	10.0°S, 22.0°E	16
Mons Pico	46.0°N, 9.0°W	29
Mons Piton	40.6°N, 1.1°W	26
Mons Rümker	40.8°N, 58.1°W	55
Mons Vinogradov	22.4°N, 32.4°W	53b
Montes Agricola	29.1°N, 54.2°W	58
Montes Alpes	46.4°N, 0.8°W	27
Montes Apenninus	18.9°N, 3.7°W	31
Montes Archimedes	25.3°N, 4.6°W	30
Montes Caucasus	38.4°N, 10.0°E	24
Montes Haemus	19.9°N, 9.2°E	32
Montes Harbinger	27.0°N, 41.0°W	58
Montes Pyrenaeus	15.6°S, 41.2°E	13
Montes Recti	48.0°N, 20.0°W	29
Montes Riphaeus	7.7°S, 28.1°W	49
Montes Spitzbergen	35.0°N, 5.0°W	30
Montes Taurus	28.4°N, 41.1°E	6
Montes Teneriffe	48.0°N, 13.0°W	29
Moretus	70.6°S, 5.8°W	41
Mösting	0.7°S, 5.9°W	35
Müller	7.6°S, 2.1°E	36
Murchison	5.1°N, 0.1°W	35
Nasireddin	41.0°S, 0.2°E	40
Nasmyth	50.5°S, 56.2°W	65
Natasha	20.0°N, 31.3°W	53b
Newcomb	29.9°N, 43.8°E	3
Newton	76.7°S, 16.9°W	41
Nicollet	21.9°S, 12.5°W	38
North Pole	90°N	57
O'Neill's Bridge	15.2°N, 49.2°E	2b
Oenopides	57.0°N, 64.1°W	56
Oken	43.7°S, 75.0°E	10
Olbers	7.4°N, 75.9°W	67
Opelt	16.3°S, 17.5°W	47
Oppolzer	1.5°S, 0.5°W	35
Orontius	40.3°S, 4.0°W	40
Pallas	5.5°N, 1.6°W	35
Palus Epidemiarum	32.0°S, 28.2°W	45
Palus Putredinis	26.5°N, 0.0°	31a
Palus Somni	14.1°N, 45.0°E	7
Parry	7.9°S, 15.8°W	48
Peary	88.6°N, 33.0°E	57
Peirce	18.3°N, 53.5°E	2b
Petavius	25.1°S, 60.4°E	9
Philolaus	72.1°N, 32.4°W	56
Phocylides	52.7°S, 57.0°W	65
Piazzi	36.5°S, 67.9°W	65
Picard	14.6°N, 54.7°E	2b
Piccolomini	29.7°S, 32.2°E	14
Pickering	2.9°S, 7.0°E	36
Pitatus	29.9°S, 13.5°W	46
Pitiscus	50.4°S, 30.9°E	11
Plana	42.2°N, 28.2°E	22
Plato	51.6°N, 9.4°W	28
Plinius	15.4°N, 23.7°E	19b
Pontécoulant	58.7°S, 66.0°E	10
Posidonius	31.8°N, 29.9°E	21
Prinz	25.5°N, 44.1°W	58
Proclus	16.1°N, 46.8°E	7
Proclus G	12.7°N, 42.7°E	7
Promontorium Kelvin	27.0°S, 33.0°W	64
Ptolemaeus	9.3°S, 1.9°W	37
Puiseux	27.8°S, 39.0°W	64
Purbach	25.5°S, 1.9°W	39
Pythagoras	63.5°N, 63.0°W	56
Pytheas	20.5°N, 20.6°W	53b
Rabbi Levi	34.7°S, 24.0°E	14
Ramsden	32.9°S, 31.8°W	45
Réaumur	2.4°S, 0.7°E	35
Regiomontanus	28.3°S, 1.0°W	39
Reiner	7.0°N, 59.0°W	61
Reiner Gamma	7.5°N, 59.0°W	61
Reinhold	3.3°N, 22.8°W	50
Reinhold B	4.3°N, 21.7°W	50
Rhaeticus	0.0°, 4.9°E	35
Rheita	37.1°S, 47.2°E	12
Riccioli	3.3°S, 74.6°W	67
Rima Ariadaeus	6.4°N, 14.0°E	33
Rima Birt	21.0°S, 9.0°W	38
Rima Bradley	23.8°N, 1.2°W	31b
Rima Calippus	37.0°N.13.0°E	24
Rima G. Bond	33.3°N, 35.5°E	21
Rima Hadley	25.0°N, 3.0°E	31b
Rima Hesiodus	31.0°S, 22.3°W	46
Rima Hyginus	7.4°N, 7.8°E	34
Rima Marius	10.0°N, 49.0°W	61
Rima Messier	1.0°S, 45.0°E	8a
Rima Schröter	1.0°N, 6.0°W	35
Rima Sharp	46.7°N, 50.5°W	55
Rima Suess	6.7°N, 48.2°W	61
Rima Sung-Mei	24.6°N, 11.3°E	20a
Rimae Archimedes	26.6°N, 4.1°W	30
Rimae Aristarchus	26.9°N, 47.5°W	58
Rimae Darwin	20.0°S, 67.0°W	66
Rimae Fresnel	28.0°N, 4.0°E	31b
Rimae Herigonius	13.0°S, 37.0°W	63
Rimae Hippalus	25.0°S, 29.0°W	64
Rimae Hypatia	1.0°S, 23.0°E	18
Rimae Mersenius	20.0°S, 45.0°W	63
Rimae Plato	51.0°N, 2.0°W	28
Rimae Secchi	1.0°N, 44.0°E	8b
Rimae Sirsalis	15.7°S, 61.7°W	66
Rimae Sulpicius Gallus	21.0°N, 10.0°E	20b
Rimae Triesnecker	4.3°N, 4.6°E	34
Ritter	2.0°N, 19.2°E	18
Rocca	12.7°S, 72.8°W	66
Römer	25.4°N, 36.4°E	6
Ross	11.7°N, 21.7°E	19b
Rosse	17.9°S, 35.0°E	13b
Rupes Altai	24.3°S, 22.6°E	14
Rupes Liebig	25.0°S, 46.0°W	64
Rupes Recta	22.1°S, 7.8°W	38
Rupes Toscanelli	27.4°N, 47.5°W	58
Russell	26.5°N, 75.4°W	60
Sabine	1.4°N, 20.1°E	18
Sacrobosco	23.7°S, 16.7°E	14
Saussure	43.4°S, 3.6°W	43
Scheiner	60.5°S, 27.5°W	42
Schiaparelli	23.4°N, 58.8°W	60
Schickard	44.3°S, 55.3°W	65
Schiller	51.9°S, 39.0°W	44
Schiller-Zucchius Basin	54.0°S, 46.0°W	44
Schlüter	5.9°S, 83.3°W	67
Schröter	2.6°N, 7.0°W	35
Scoresby	77.7°N, 14.1°E	57
Secchi	2.4°N, 43.5°E	8b
Segner	58.9°S, 48.3°W	44
Seleucus	21.0°N, 66.6°W	60
Sharp	45.7°N, 40.2°W	54
Sinus Aestuum	10.9°N, 8.8°W	52
Sinus Asperatitis	6.0°S, 25.0°E	17
Sinus Concordiae	10.8°N, 43.2°E	7
Sinus Iridum	45.0°N, 32.0°W	54
Sinus Lunicus	32.0°N, 1.0°W	25
Sinus Medii	0.0°, 0.0°	35
Sirsalis	12.5°S, 60.4°W	66
Sosigenes	8.7°N, 17.6°E	19b
South	58.0°N, 50.8°W	56
Stadius	10.5°N, 13.7°W	52
Statio Tranquillitatis	0.7°N, 23.5°E	18
Steinheil	48.6°S, 46.5°E	12
Stevinus	32.5°S, 54.2°E	9
Stöfler	41.1°S, 6.0°E	40
Strabo	61.9°N, 54.3°E	4
Struve	22.4°N, 77.1°W	60
South Pole	90°S	41
Sulpicius Gallus	19.6°N, 11.6°E	20b
Sulpicius Gallus Formation	20.0°N, 10.0°E	20b
Taruntius	5.6°N, 46.5°E	7
Tempel	3.9°N, 11.9°E	33
Thales	61.8°N, 50.3°E	4
The Helmet	16.7°S, 31.5°W	63
Theaetetus	37.0°N, 6.0°E	26
Thebit	22.0°S, 4.0°W	38
Theophilus	11.4°S, 26.4°E	16
Timocharis	26.7°N, 13.1°W	53b
Tolansky	9.5°S, 16.0°W	48
Torricelli	4.6°S, 28.5°E	17
Turner	1.4°S, 13.2°W	48
Tycho	43.4°S, 11.1°W	43
Valentine Dome	31.0°N, 10.3°E	24
Vallis Alpes	48.5°N, 3.2°E	27
Vallis Rheita	42.5°S, 51.5°E	12
Vallis Schröteri	26.2°N, 50.8°W	58
Vasco da Gama	13.6°N, 83.9°W	60
Vendelinus	16.4°S, 61.6°E	9
Vitello	30.4°S, 37.5°W	64
Vlacq	53.3°S, 38.8°E	11
Vogel	15.1°S, 5.9°E	36
W. Bond	65.4°N, 4.5°E	57
Wallace	20.3°N, 8.7°W	31a
Walther	33.1°S, 1.0°E	39
Wargentin	49.6°S, 60.2°W	65
Watt	49.5°S, 48.6°E	12
Werner	28.0°S, 3.3°E	39
Wichmann	7.5°S, 38.1°W	62
Wilhelm	43.4°S, 20.4°W	43
Winthorp	10.7°S, 44.4°W	62
Wolf	22.7°S, 16.6°W	47
Yerkes	4.6°N, 51.7°E	2b
Zagut	32.0°S, 22.1°E	14
Zucchius	61.4°S, 50.3°W	44
Zupus	17.2°S, 53.0°W	62
Apollo 11	0.7°N, 23.5°E	2, 18, 19
Apollo 12	3.0°S, 23.2°W	51
Apollo 14	3.6°S, 17.4°W	48
Apollo 15	26.0°N, 3.6°E	20, 31, 58
Apollo 16	9.0°S, 15.6°E	15, 63
Apollo 17	20.1°N, 30.8°E	6, 20
Surveyor 1	2.5°S, 43.2°W	62
Surveyor 4	0.4°N, 1.3°W	35
Surveyor 5	1.4°N, 23.2°E	19
Surveyor 6	0.4°N, 1.4°W	35
Ranger 6	9.3°N, 21.5°E	19
Ranger 7	10.7°S, 20.6°W	37, 49
Ranger 8	2.8°N, 24.9°E	19
Luna 1	–	25
Luna 2	30.1°N, 0.1°W	25
Luna 3	–	68
Luna 7	9.8°N, 47.8°W	61
Luna 15	11.7°N, 58.5°E	2
Luna 16	0.8°N, 56.4°E	2
Luna 20	3.4°N, 56.4°E	2
Luna 21	25.9°N, 30.3°E	20
Luna 24	14.0°N, 66.0°E	2

Image credits

M. Clark, Whitepeak Lunar Observatory: 28 all (4×)
K. Gilbert: 14
M. Levens/Volkssternwarte Hannover: 27 left
NASA: 6, 8 both, 9 all, 11, 16 above, 17 centre right, 19 right
(3×) and left below, 20 below left (2×), 21 all (4×), 22 all (2×),
23 above left and below right, 25 above left
W. Paech: front endpapers (3×), 5, 10, 12, 13 both, 15 below
(3×), 16 below (3×), 18 all (3×), 23 centre left, 25 below right,
26 above, 27 right, 29, 30 all (2×), 31 all (2×), rear endpapers
(2×)
M. Rietze: 17 centre left, 17 below right, 20 above right
W. Sorgenfrey: 27 right
M. Theusner: 7 above, 17 above, 24

*The images on the atlas pages are sorted by section and by
photographers Alan Chu (AC), Wolfgang Paech (WP), Wolfgang Sorgenfrey (WS), Michael Theusner (MT) and Mario
Weigand (MW).*
*The abbreviations show the image location as follows: 1 =
first double-page spread, 2 = second double-page spread, l =
left side of the spread, r = right side of the spread, a = above,
c = centre, b = below*

1-Mare Smythii
AC: lb, rcb / WP: la, rb / MT: ra

2-Mare Crisium
AC: 2la, 2lb, 2ra / WP: 1l, / MW: 1r, 2la, 2c, 2rb

3-Cleomedes
AC: rc, rb / WP: lb, la, ra / MW: rcb

4-Endymion
AC: lc / WP: la, rca, ra, lb / MW: rb

5-Atlas/Hercules
WP: la, lb, ra / MW: rc, rb

6-Montes Taurus
WP: all

7-Palus Somni
AC: rc / WP: l, r / MW: rca, rcb

8-Mare Fecunditatis
AC: 2rc, 2rb / WP: 1l, 1rca, 1rcb, 2l, 2lca, 2lcb, 2ra / MW: 1rb

9-Langrenus/Petavius
AC: rcb, rb / MW: l, ra, rca, ra / WS: rc

10-Mare Australe
AC: la / WP: lb, rca / MT: rb

11-Vlacq
AC: rb / WP: la, lb, ra / MW: rc

12-Vallis Rheita
AC: rb / WP: la, ra / MW: lca, rc

13-Mare Nectaris
AC: 2rcb, 2rb / WP: 1l, 1rb, 1ra, 1rca, 2la, 2ra / WS: 2lc

14-Rupes Altai
AC: ra / WP: la, lb, lbc, rb

15-Abulfeda
WP: all

16-Theophilus
WP: la, lcb, lb, ra / MW: r, rb

17-Sinus Asperitatis
AC: lb, lcb / WP: la, lca, rca, ra, rb

18-Statio Tranquillitatis
NASA: lb / WP: la / MW: ra, rc, rb

19-Mare Tranquillitatis
AC: 2lcb, 2rc / WP: 1l, 1ra, 2la, 2ra, 2rca / WS: 1rb / MW: 2lca,
2lb, 2rb

20-Mare Serenitatis
AC: 1rcb, 2rc / WP: 1la, 1lb, 1ra, 1rca, 1rb, 2lb, 2lca / WS: 2la

/ MW: 2ra

21-Posidonius
AC: rb / WP: lb, lc / MW: la, ra, rc

22-Lacus Mortis
WP: la, lca, lcb, rcb, rb / MW: rca, ra, rc

23-Aristoteles/Eudoxus
WP: la, lb, lcb / MW: ra, rcb, rb

24-Montes Caucasus
AC: ra / WP: la, lb, rb

25-Autolycus/Aristillus
WP: ra, rb / MW: la, lb, lcb

26-Cassini
WP: lb, lcb / MW: ra, rb

27-Montes Alpes
WP: la, lb, ra, rcb, rb / MW: rca

28-Plato
AC: lc / WP: la, ra / MW: b

29-Montes Teneriffe
WP: la, lb, lcb / MW: ra, rb, rcb

30-Archimedes
WP: all

31-Montes Apenninus
AC: 2rb / WP: 1all, 2la, lca / MW: 2lb, 2ra

32-Mare Vaporum
AC: rcb / WP: l / MW: ra, rb

33-Rima Ariadaeus
AC: rc, rcb, rb / WP: lb, la, ra

34-Rima Hyginus
WP: la, lb, rcb / MW: ra, rb

35-Sinus Medii
AC: lb / WP: la, ra, rb

36-Hipparchus
WP: all

37-Ptolemaeus
AC: 1rc, 1rb / WP: 1l, 1lc, 1ra, 1rcb, 2la, 2lb / MW: 2lcl, 2lc, 2r

38-Rupes Recta
WP: ra, rb / MW: la, lb

39-Regiomontanus
WP: la, lc, ra / MW: rb

40-Maurolycus
WP: la, lc, ra, rcb, rb / MW: lb

41-South Pole
WP: la, ra, rb / MW: lc, lb, rc

42-Clavius
WP: la / MW: ra, c

43-Tycho
AC: rb / WP: lb, rc / MT: la / MW: ra, rca, rcb

44-Schiller
AC: lm, lb, rb / WP: c / MW: lcb, ra

45-Palus Epidemiarum
AC: rc / WP: l, ra, rcl / MW: rb

46-Pitatus
AC: rb / WP: lb, la / MT: ra

47-Mare Nubium
AC: rb / WP: rc, ra / MT: l

48-Fra Mauro
AC: lc, rcb, rb / WP: ra / MW: lb

49-Mare Cognitum
AC: lb, rb / WP: la, ra

50-Mare Insularum
AC: ra, rc / WP: la, rb / WS: lb

51-Copernicus
AC: 1rc / WP: 1la, 1lb, 2all / MT: 1rcb / MW: 1rca, 1ra, 1rb

52-Eratosthenes
WP: la, lb, rca, rcb, rb / MW: ra

53-Mare Imbrium
AC: 2rb / WP: 2la, 2lb, 2ra / MW: 1, 2rc

54-Sinus Iridum
WP: ra, rb / MW: l

55-Gruithuisen
AC: lcb / NASA: ra / WP: la / MW: lb, rc

56-Mare Frigoris
AC: lb / WP: rc / MW: o, rb

57-North Pole
WP: all

58-Aristarchus
AC: ra / WP: rb / MW: l, rc

59-Kepler
WP: la, ra, rcb / MW: lb, lcb, rb

60-Seleucus
AC: ra, rb / WP: la, lb / MW: rcb

61-Reiner
AC: la, rb / WP: lc / MW: ra

62-Letronne/Hansteen
AC: lcb, rc / WP: la / WS: lb / MW: ra, rb

63-Gassendi
AC: lb, lcb / WP: la / MW: rca, ra, rb

64-Mare Humorum
AC: ra / WP: lb / MT: la / MW: rc

65-Schickard
AC: rb, rcb / MW: l, ra

66-Sirsalis
AC: all

67-Grimaldi
AC: lcr, rb / WP: la, lc / WS: ra / MW: lcl

68-Mare Orientale
NASA: lb / WP: la

69-Lunar Farside
NASA: r

Further reading and references

Books

Bussey, B., Spudis, P. D.: The Clementine Atlas of the Moon, Cambridge University Press, 2004 (Revised edition 2012)

Byrne, C. J.: Lunar Orbiter Photographic Atlas of the Near Side of the Moon, Springer-Verlag 2005

Grego, P.: The Moon and How to Observe It, Springer-Verlag 2010

Light, M.: Full Moon, Jonathan Cape, 1999

Mailer, N.: Moonfire: The Epic Journey of Apollo 11, Taschen-Verlag 2010

North, G., Observing the Moon, Cambridge University Press, 2000 (Second edition 2007)

Rükl, A.: Atlas of the Moon, Sky Publishing Corp, 2004

Stooke, P. J.: The International Atlas of Lunar Exploration, Cambridge University Press, 2007

Whitaker, E. A., Mapping and Naming the Moon, Cambridge University Press, 1999

Wood, C. A.: The Modern Moon, A Personal View, Sky Publishing 2007

Software

Lunar Terminator Visualization Tool: http://ltvt.wikispaces.com/LTVT

MoonZoo: www.moonzoo.org

Virtual Moon Atlas: www.astrosurf.com/avl/UK_download.html

CD

Consolidated Lunar Atlas: 2 CDs, Lunar and Planetary Institute, 2003

Internet links

ALPO Lunar Section: http://alpo-astronomy.org/lunarblog

Apollo Images: www.hq.nasa.gov/alsj/picture.html

The Astronomy League (US) Lunar Program: www.astroleague.org/al/obsclubs/lunar/lunar1.html

BAA Lunar Section: www.baalunarsection.org.uk

Catalogue of Lunar Domes: http://digilander.libero.it/glrgroup/consolidatedlunardomecatalogue.htm

Consolidated Lunar Atlas: www.lpi.usra.edu/resources/cla

Digital Lunar Orbiter Atlas: www.lpi.usra.edu/resources/lunar_orbiter

Geological maps: www.lpi.usra.edu/resources/mapcatalog/usgs/

List of lunar nomenclature: www.planet4589.org/astro/lunar/

Lunar crater rays: www.lunar-occultations.com/rlo/rays/rays.htm

Lunar missions: www.lpi.usra.edu/expmoon/

Lunar Picture of the Day: http://lpod.wikispaces.com

NASA list of links: http://nssdc.gsfc.nasa.gov/planetary/planets/moonpage.html

NSSDC Lunar Image Catalog: http://nssdc.gsfc.nasa.gov/imgcat/html/group_page/EM.html

Society for Popular Astronomy (UK) Lunar Section: www.popastro.com/lunar/index.php

USGS Gazetteer of Lunar Nomenclature: http://planetarynames.wr.usgs.gov/Page/MOON/target

Authors

Alan Chu: www.alanchuhk.com

Wolfgang Paech: www.astrotech-hannover.de

Mario Weigand: www.skytrip.de

Wolfgang Sorgenfrey: www.sorgenfreyfotografien.de

Michael Theusner: www.theusner.eu

Key to Sections 41–69

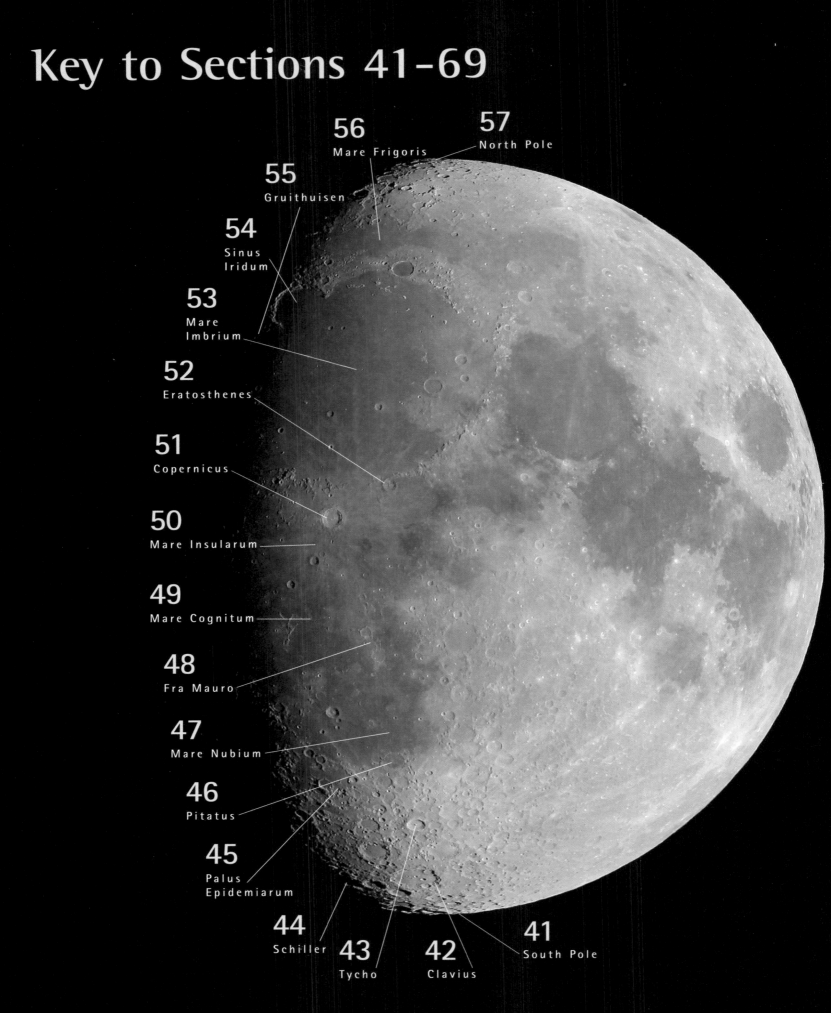

56 Mare Frigoris

57 North Pole

55 Gruithuisen

54 Sinus Iridum

53 Mare Imbrium

52 Eratosthenes

51 Copernicus

50 Mare Insularum

49 Mare Cognitum

48 Fra Mauro

47 Mare Nubium

46 Pitatus

45 Palus Epidemiarum

44 Schiller

43 Tycho

42 Clavius

41 South Pole

The key to Sections 1–40 is shown on the front endpaper.